MICROGRIDS

MICROGRIDS

ARCHITECTURES AND CONTROL

Edited by
Professor Nikos Hatziargyriou
National Technical University of Athens, Greece

This edition first published 2014
© 2014 John Wiley and Sons Ltd

Registered office

John Wiley & Sons Ltd, The Atrium, Southern Gate, Chichester, West Sussex, PO19 8SQ, United Kingdom

For details of our global editorial offices, for customer services and for information about how to apply for permission to reuse the copyright material in this book please see our website at www.wiley.com.

Library of Congress Cataloging-in-Publication Data

Microgrid : architectures and control / edited by professor Nikos Hatziargyriou.
     1 online resource.
  Includes bibliographical references and index.
  Description based on print version record and CIP data provided by publisher; resource not viewed.
   ISBN 978-1-118-72064-6 (ePub) – ISBN 978-1-118-72065-3 – ISBN 978-1-118-72068-4 (cloth)
1. Smart power grids.   2. Small power production facilities.   I. Hatziargyriou, Nikos, editor of compilation.
  TK3105
  621.31–dc23

                                           2013025351

A catalogue record for this book is available from the British Library.

ISBN: 978-1-118-72068-4

Set in 10/12 pt Times by Thomson Digital, Noida, India

Dedicated to the Muse of Creativity

Contents

Foreword

The idea of microgrids is not new. However, as new technologies are coming into existence to harvest renewable energy as well as more efficient electricity production methods coupled with the flexibility of power electronics; a new industry is developing to promote these technologies and organize them into microgrids for extracting the maximum benefits for owners and the power grid. More than 15 years ago, the Department of Energy has sponsored early research that laid the foundations for microgrids and explored the benefits. One key aspect is the ability and promise to address environmental concerns that have been growing in recent years. Today the microgrid concept has exploded to include a variety of architectures of energy resources into a coordinated energy entity that its value is much greater than the individual components. As a result the complexity of microgrids has increased. It is in this environment of evolution of microgrids that the present book is very welcome. It is written in a way that provides valuable information for specialist as well as non-specialists.

Chapter 1 provides a well thought view of the microgrid concept from the various forms of implementation to the potential economic, environmental and technical benefits. It identifies the role of microgrids in altering the distribution system as we know it today and at the same time elaborates on the formation of microgrids as an organized entity interfaced to distribution systems. In a refreshingly simple way identifies the enabling technologies for microgrids, that is power electronics, communications, renewable resources. It discusses in simple terms the ability of microgrids to minimize green house gases, help the power grid with load balancing and voltage control and assist power markets. While it is recognized that participation of the microgrids in power markets is limited by their size, it discusses possible ways that microgrids can market their assets via aggregators and opens the field for other innovations.

The book addresses two of the great challenges of microgrids: control and protection. Four chapters are devoted to these complex problems, three on control (Chapters 2, 3 and 5) and one on protection (Chapter 4). The multiplicity of control issues and their complexity is elaborated in a clear and concise manner. Since microgrids comprise many resources that are interfaced via power electronics, the book presents the organization of the control problems in a hierarchical architecture that consists of local controllers that control specific resources, their operation and their protection as well as outer loop controllers that perform load-generation management, islanding operation as well as the interaction with up-stream controllers including power system control centers. It provides a good overview of approaches as well as the role of state estimation in controlling and operating a microgrid. In addition to conventional control methods, recent intelligent control approaches are also discussed. The specific issues and challenges of microgrid control are clearly elaborated. As an example, because the microgrid typically comprises many inverters connecting various resources to the microgrid it is possible to trigger oscillations due to inverter control interactions. Methods for solving these issues are clearly discussed in an easy to follow way. It is recognized that multiple microgrids can exist in a system and the issue of controlling and coordinating all the microgrids is very important from the point of view of managing the microsources as well as providing services to the power grid by coordinating all the resources. The services can be any of the ancillary services that are typically provided by large systems: frequency control, voltage control, power balance, capacity reserves. The hierarchies involved in the control and operation of multi-microgrid systems is eloquently presented as a hierarchical control problem.

Protection of microgrids is a challenging problem due to the fact that microgrid resources provide limited fault currents. Detection of faults in microgrids is problematic at best because the grid side fault current contribution may be very high while the contribution from microsources is limited. Present protection schemes and functions are not reliable for microgrids. The book describes clever methods for providing adequate protection functions such as adaptive protection schemes, addition of components that will provide temporarily high fault currents to enable the operation of protective relays, increasing inverter capacity and therefore fault current contribution. While the book provides some solutions it also makes it clear that there is much more work that needs to be done to reliably protect microgrids.

The basic approaches in designing, controlling and protecting microgrids are nicely complimented by a long list of microgrid projects around the globe that provide a picture of the evolution of microgrid design and lessons learned. Specific microgrid projects in Europe, United States, Japan, China and Chile are described and discussed. These projects provide an amazing insight into the lessons learned, challenges faced and issues resolved and issues outstanding. The examples span small capacity microgrids as well as some very large microgrids; grid-connected microgrids as well as stand-alone or island microgrids. The information provided is extremely useful and enables appreciation of the challenges as well as the rewards of these systems.

Finally, the last chapter elaborates on the technical, economic, environmental and social benefits of microgrids. The discussion is qualitative as well as quantitative. While the quantitative analysis is very much dependent upon specific areas and other conditions, the qualitative discussion is applicable to microgrids anywhere in the globe. Indirectly, this discussion makes the case for microgrids comprising mostly renewable energy resources as a big component in solving the environmental, economic and social issues that are facing a society that relies more and more in electric energy. The technical issues are solvable for transforming distribution systems into a distributed microgrid. The work presented in this book will be a fundamental reference toward the promotion and proliferation of microgrids and the accompanied deployment of renewable resources.

This book is a must read resource for anyone interested in the design and operation of microgrids and the integration of renewable resources into the power grid.

Sakis Meliopoulos
Georgia Power Distinguished Professor
School of Electrical and Computer Engineering
Georgia Institute of Technology
Atlanta, Georgia

Preface

The book deals with understanding, analyzing and justifying Microgrids, as novel distribution network structures that unlock the full potential of Distributed Energy Resources (DER) and thus form building blocks of future Smartgrids. In the context of this book, Microgrids are defined as distribution systems with distributed energy sources, storage devices and controllable loads, operated connected to the main power network or islanded, in a controlled, coordinated way. Coordination and control of DER is the key feature that distinguishes Microgrids from simple distribution feeders with DER. In particular, effective energy management within Microgrids is the key to achieving vital efficiency benefits by optimizing production and consumption of energy. Nevertheless, the technical challenges associated with the design, operation and control of Microgrids are immense. Equally important is the economic justification of Microgrids considering current electricity market environments and the quantified assessment of their benefits from the view of the various stakeholders involved.

Discussions about Microgrids started in the early 2000, although their benefits for island and remote, off-grid systems were already generally appreciated. Nowadays, Microgrids are proposed as vital solutions for critical infrastructures, campuses, remote communities, military applications, utilities and communal networks. Bright prospects for a steady market growth are foreseen. The book is intended to meet the needs of practicing engineers, familiar with medium- and low-voltage distribution systems, utility operators, power systems researchers and academics. It can also serve as a useful reference for system planners and operators, technology providers, manufacturers and network operators, government regulators, and postgraduate power systems students.

The text presents results from a 6-year joint European collaborative work conducted in the framework of two EC-funded research projects. These are the projects "Microgrids: *Large Scale Integration of Micro-Generation to Low Voltage Grids*," funded within the 5th Framework programme (1998–2002) and the follow-up project "More Microgrids, Advanced Architectures and Control Concepts for More Microgrids" funded within the 6th Framework Programme (2002–2006). The consortia involved were coordinated by the editor of this book and comprised a number of industrial partners, power utilities and academic research teams from 12 EU countries. A wealth of information and many practical conclusions were derived from these two major research efforts. The book attempts to clarify the role of Microgrids within the overall power system structure and focuses on the main findings related to primary and secondary control and management at the Microgrid and Multi-Microgrid level. It also provides results from quantified assessment of the Microgrids benefits from an economical, environmental, operational and social point of view. A separate chapter beyond the EC projects, provided by a more international authorship is devoted to an overview of real-world Microgrids from various parts of the world, including, next to Europe, United States of America, Japan, China and Chile.

Chapter 1, entitled "The Microgrids Concept," co-authored by Christine Schwaegerl and Liang Tao, clarifies the key features of Microgrids and underlines the distinguishing characteristics from other DG dominated structures, such as Virtual Power Plants. It discusses the main features related to their operation and control, the market models and the effect of possible regulatory settings and provides an exemplary roadmap for Microgrid development in Europe.

Chapter 2, entitled "Microgrids Control Issues" co-authored by Aris Dimeas, Antonis Tsikalakis, George Kariniotakis and George Korres, deals with one of the key features of Microgrids, namely their energy management. It presents the hierarchical control levels distinguished in Microgrids operation and discusses the principles and main functions of centralized and decentralized control, including forecasting and state estimation. Next, centralized control functions are analyzed and illustrated by a practical numerical example. Finally, an overview of the basic multi-agent systems concepts and their application for decentralized control of Microgrids is provided.

Chapter 3, entitled "Intelligent Local Controllers," co-authored by Thomas Degner, Nikos Soultanis, Alfred Engler and Asier Gil de Muro, presents primary control capabilities of DER controllers. The provision of ancillary services in interconnected mode and the capabilities of voltage and frequency control, in case of islanded operation and during transition between the two modes are outlined. Emphasis is placed on the implications of the high resistance over reactance ratios, typically found in LV Microgrids. A control algorithm based on the fictitious impedance method to overcome the related problems together with characteristic simulation results are provided.

Chapter 4, entitled "Microgrid Protection," co-authored by Alexander Oudalov, Thomas Degner, Frank van Overbeeke and Jose Miguel Yarza, deals with methods for effective protection in Microgrids. A number of challenges are caused by DER varying operating conditions, the reduced fault contribution by power electronics interfaced DER and the occasionally increased fault levels. Two adaptive protection techniques, based on pre-calculated and on-line calculated settings are proposed including practical implementation issues. Techniques to increase the amount of fault current level by a dedicated device and the possible use of fault current limitation are also discussed.

Chapter 5, entitled "Operation of Multi-Microgrids," co-authored by João Abel Peças Lopes, André Madureira, Nuno Gil and Fernanda Resende examines the operation of distribution networks with increasing penetration of several low voltage Microgrids, coordinated with generators and flexible loads connected at medium voltage. An hierarchical management architecture is proposed and functions for coordinated voltage/VAR control and coordinated frequency control are analyzed and simulated using realistic distribution networks. The capability of Microgrids to provide black start services are used to provide restoration guidelines. Finally, methods for deriving Microgrids equivalents for dynamic studies are discussed.

Chapter 6, entitled "Pilot Sites: Success Stories and Learnt Lessons" provides an overview of real-world Microgrids, already in operation as off-grid applications, pilot cases or full-scale demonstrations. The material is organized according to geographical divisions. George Kariniotakis, Aris Dimeas and Frank van Overbeeke describe three pilot sites in Europe developed within the more Microgrids project; John Romankiewicz and Chris Marnay provide an overview of Microgrid Projects in the United States; Satoshi Morozumi provide an overview of the Japanese Microgrid Projects; Meiqin Mao describes the Microgrid Projects in China and Rodrigo Palma Behnke and Guillermo Jiménez-Estévez provide details of an off-grid Microgrid in Chile. These projects are of course indicative of a continuously growing list, they provide, however, a good impression of the on-going developments in the field.

Chapter 7, entitled "Quantification of Technical, Economic, Environmental and Social Benefits of Microgrid Operation," co-authored by Christine Schwaegerl and Liang Tao attempts to quantify the Microgrids benefits using typical European distribution networks of different types and assuming various DER penetration scenarios, market conditions, prices and costs developments for the years 2020, 2030 and 2040. Sensitivity analysis of the calculated benefits is performed. Although, the precision of these quantified benefits is subject to the high uncertainties in the underlying assumptions, the positive effects of Microgrids operation can be safely observed in all cases.

Next to the co-authors of the various chapters, there are many researchers who have contributed to the material of this book by their knowledge, research efforts and fruitful collaboration during the numerous technical meetings of the Microgrids projects. I am indebted to all of them, but I feel obliged to refer to some names individually and apologize in advance for the names I might forget. I would like to start with Profs. Nick Jenkins and Goran Strbac from UK; I have benefited tremendously while working with them and their insights and discussions helped clarify many concepts discussed in the book. I am indebted to Britta Buchholz, Christian Hardt, Roland Pickhan, Mariam Khattabi, Michel Vandenbergh, Martin Braun, Dominik Geibel and Boris Valov from Germany; Mikes Barnes, Olimpo Anaya-Lara, Janaka Ekanayake, Pierluigi Mancarella, Danny Pudjianto and Tony Lakin from UK; Jose Maria Oyarzabal, Joseba Jimeno and Iñigo Cobelo from Spain; Nuno Melo and António Amorim from Portugal; Sjef Cobben from the Netherlands; John Eli Nielsen from Denmark; Perego Omar and Michelangeli Chiara from Italy; Aleksandra Krkoleva, Natasa Markovska and Ivan Kungulovski from FYR of Macedonia; Grzegorz Jagoda and Jerszy Zielinski from Poland; my NTUA colleagues Stavros Papathanassiou and Evangelos Dialynas; and Stathis Tselepis, Kostas Elmasides, Fotis Psomadellis, Iliana Papadogoula, Manolis Voumvoulakis, Anestis Anastasiadis, Fotis Kanellos, Spyros Chadjivassiliadis and Maria Lorentzou from Greece. I express my gratitude to my PhD students and collaborators Georgia Asimakopoulou, John Karakitsios, Evangelos Karfopoulos, Vassilis Kleftakis, Panos Kotsampopoulos, Despina Koukoula, Jason Kouveliotis-Lysicatos, Alexandros Rigas, Nassos Vassilakis, Panayiotis Moutis, Christina Papadimitriou and Dimitris Trakas, who reviewed various chapters of the book and provided valuable comments. Finally, I wish to thank the EC DG Research&Innovation for providing the much appreciated funding for the research leading to this book, especially the Officers Manuel Sanchez Jimenez and Patrick Van Hove.

Nikos Hatziargyriou

List of Contributors

Thomas Degner, Thomas Degner is Head of Department Network Technology and Integration at Fraunhofer IWES, Kassel, Germany. He received his Diploma in Physics and his Ph.D. from University of Oldenburg. His particular interests include microgrids, interconnection requirements and testing procedures for distributed generators, as well as power system stability and control for island and interconnected power systems with a large share of renewable generation.

Aris Dimeas, Aris L. Dimeas received the Diploma and his Ph.D. in Electrical and Computer Engineering from the National Technical University of Athens (NTUA). He is currently senior researcher at the Electrical and Computer Engineering School of NTUA. His research interests include dispersed generation, artificial intelligence techniques in power systems and computer applications in liberalized energy markets.

Alfred Engler, Alfred Engler received his Dipl.-Ing. (Master's) from the Technical University of Braunschweig and the degree Dr.-Ing. (Ph.D.) from the University of Kassel. He has been head of the group *Electricity Grids* and of *Power Electronics* at ISET e.V. involved with inverter control, island grids, microgrids, power quality and grid integration of wind power. He is currently Manager of the Department of Advance Development at Liebherr Elektronik GmbH.

Nuno Gil, Nuno José Gil received his electrical engineering degree from the University of Coimbra and M.Sc. and Ph.D. from the University of Porto. He is a researcher in the Power Systems Unit of INESC Porto and assistant professor at the Polytechnic Institute of Leiria, Portugal. His research interests include integration of distributed generation and storage devices in distribution grids, islanded operation and frequency control.

Guillermo Jiménez-Estévez, Guillermo A. Jiménez-Estévez received the B.Sc. degree in Electrical Engineering from the Escuela Colombiana de Ingeniería, Bogotá, and the M.Sc. and Ph.D. degrees from the University of Chile, Santiago. He is assistant director of the Energy Center, FCFM, University of Chile.

George Kariniotakis, Georges Kariniotakis received his engineering and M.Sc. degrees from the Technical University of Crete, Greece and his Ph.D. degree from Ecole des Mines de Paris. He is currently with the Centre for Processes, Renewable Energies & Energy Systems (PERSEE) of MINES ParisTech as senior scientist and head of the Renewable Energies & Smartgrids Group. His research interests include renewables, forecasting and smartgrids.

George Korres, George N. Korres received the Diploma and Ph.D. degrees in Electrical and Computer Engineering from the National Technical University of Athens. He is professor with the School of Electrical and Computer Engineering of NTUA. His research interests are in power system state estimation, power system protection and industrial automation.

André Madureira, André G. Madureira received an Electrical Engineering degree, M.Sc. and Ph.D. from the Faculty of Engineering of the University of Porto. He is senior researcher/consultant in the Power Systems Unit of INESC Porto. His research interests include integration of distributed generation and microgeneration in distribution grids, voltage and frequency control and smartgrid deployment.

Meiqin Mao, Meiqin Mao is professor with Research Center of Photovoltaic System Engineering, Ministry of Education, Hefei University of Technology, P.R. China. She has been devoted to renewable energy generation research since 1993. Her research interests include optimal operation and energy management of microgrids and power electronics applications in renewable energy systems.

Chris Marnay, Chris Marnay has been involved in microgrid research since the 1990s. He studies the economic and environmental optimization of microgrid equipment selection and operation. He has chaired nine of the annual international microgrid symposiums.

Satoshi Morozumi, Satoshi Morozumi has graduated from a doctor course in Hokkaido University. He joined Mitsubishi Research Institute, Inc. and for past 20 years, he was engaged in the utility's system. He is currently director general of Smart Community Department at NEDO, where he is in charge of management of international smart community demonstrations.

Asier Gil de Muro, Asier Gil de Muro has received his M.Sc. in Electrical Engineering from the School of Engineering of the University of the Basque Country in Bilbao, Spain. Since 1999 he is researcher and project manager at the Energy Unit of TECNALIA, working in projects dealing with power electronics equipment, design and developing of grid interconnected power devices, microgrids and active distribution.

Alexandre Oudalov, Alexandre Oudalov received the Ph.D. in Electrical Engineering in 2003 from the Swiss Federal Institute of Technology in Lausanne (EPFL), Switzerland. He joined ABB Switzerland Ltd., Corporate Research Center in 2004 where he is currently a principal scientist in the Utility Solutions group. His research interests include T&D grid automation, integration and management of DER and control and protection of microgrids

Frank van Overbeeke, Frank van Overbeeke graduated in Electrical Power Engineering at Delft University of Technology and obtained a Ph.D. in Applied Physics from the University of Twente. He is founder and owner of EMforce, a consultancy firm specialized in power electronic applications for distribution networks. He has acted as the system architect for several major utility energy storage projects in the Netherlands.

Rodrigo Palma Behnke, Rodrigo Palma Behnke received his B.Sc. and M.Sc. on Electrical Engineering from the Pontificia Universidad Católica de Chile and a Dr.-Ing. from the University of Dortmund, Germany. He is the director of the Energy Center, FCFM, and the Solar Energy Research Center SERC-Chile, he also is associate professor at the Electrical Engineering Department, University of Chile.

João Abel Peças Lopes, João Abel Peças Lopes is full professor at Faculty of Engineering of University of Porto and member of the Board of Directors of INESC Porto

Fernanda Resende, Fernanda O. Resende received an Electrical Engineering degree, from the University of Trás-os-Montes e Alto Douro and M.Sc. and Ph.D. from the Faculty of Engineering of the University of Porto. She is senior researcher in the Power Systems Unit of INESC Porto and assistant professor at Losófona University of Porto. Her research interests include integration of distributed generation, modeling and control of power systems and small signal stability.

John Romankiewicz, John Romankiewicz is a senior research associate at Berkeley Lab, focusing on distributed generation and energy efficiency policy. He has been working in the energy sector in the United States and China for the past 7 years and is currently studying for his Masters of Energy and Resources and Masters of Public Policy at UC Berkeley.

Christine Schwaegerl, Christine Schwaegerl received her Diploma in Electrical Engineering at the University of Erlangen and her Ph.D. from Dresden Technical University, Germany. In 2000, she joined Siemens AG where she has been responsible for several national and international research and development activities on power transmission and distribution networks. Since 2011 she is professor at Augsburg University of Applied Science.

Nikos Soultanis, Nikos Soultanis graduated from the Electrical Engineering Department of the National Technical University of Athens (NTUA) in 1989 and has worked on various projects as an independent consultant. He received his M.Sc. from the University of Manchester Institute of Science and Technology (UMIST) and his Ph.D. from NTUA. He works currently in the dispatching center of the Greek TSO and as an associated researcher at NTUA with interests in the area of distributed generation applications and power system analysis.

Liang Tao, Liang Tao is a technical consultant working at Siemens PTI, Germany. His main research interests include stochastic modeling of renewable energy sources, dimensioning and operation methods for storage devices, optimal power flow and optimal scheduling of multiple types of resources in smartgrids.

Antonis Tsikalakis, Antonis G. Tsikalakis received the Diploma and Ph.D. degrees in electrical and computer engineering from NTUA. Currently he is an adjunct lecturer in the School of Electronics and Computer Engineering of the Technical University of Crete (TUC) and research associate of the Technological Educational Institute of Crete. He co-operates with the NTUA as post-doc researcher. His research interests are in distributed generation, energy storage and autonomous power systems operation with increased RES penetration.

José Miguel Yarza, José Miguel Yarza received an M.S. degree in Electrical Engineering and a master degree on "Quality and Security in Electrical Energy Delivery. Power System Protections" from the University of Basque Country and an executive MBA from ESEUNE Business School. He is currently CTO at CG Automation. He is member of AENOR SC57 (Spanish standardization body), IEC TC57 and CIGRE.

1

The Microgrids Concept

Christine Schwaegerl and Liang Tao

1.1 Introduction

Modern society depends critically on a secure supply of energy. Growing concerns for primary energy availability and aging infrastructure of current electrical transmission and distribution networks are increasingly challenging security, reliability and quality of power supply. Very significant amounts of investment will be required to develop and renew these infrastructures, while the most efficient way to meet social demands is to incorporate innovative solutions, technologies and grid architectures. According to the International Energy Agency, global investments required in the energy sector over the period 2003–2030 are estimated at $16 trillion.

Future electricity grids have to cope with changes in technology, in the values of society, in the environment and in economy [1]. Thus, system security, operation safety, environmental protection, power quality, cost of supply and energy efficiency need to be examined in new ways in response to changing requirements in a liberalized market environment. Technologies should also demonstrate reliability, sustainability and cost effectiveness. The notion of **smart grids** refers to the evolution of electricity grids. According to the European Technology Platform of Smart Grids [2], a smart grid is an electricity network that can intelligently integrate the actions of all users connected to it – generators, consumers and those that assume both roles – in order to efficiently deliver sustainable, economic and secure electricity supplies. A smart grid employs innovative products and services together with intelligent monitoring, control, communication and self-healing technologies.

It is worth noting that power systems have always been "smart", especially at the transmission level. The distribution level, however, is now experiencing an evolution that needs more "smartness", in order to

- facilitate access to distributed generation [3,4] on a high share, based on renewable energy sources (RESs), either self-dispatched or dispatched by local distribution system operators

---

Microgrids: Architectures and Control, First Edition. Edited by Nikos Hatziargyriou.
© 2014 John Wiley & Sons, Ltd. Published 2014 by John Wiley & Sons, Ltd.
Companion Website: www.wiley.com/go/hatziargyriou_microgrids

- enable local energy demand management, interacting with end-users through smart metering systems
- benefit from technologies already applied in transmission grids, such as dynamic control techniques, so as to offer a higher overall level of power security, quality and reliability.

In summary, distribution grids are being transformed from **passive** to **active networks**, in the sense that decision-making and control are distributed, and power flows bidirectional. This type of network eases the integration of DG, RES, demand side integration (DSI) and energy storage technologies, and creates opportunities for novel types of equipment and services, all of which would need to conform to common protocols and standards. The main function of an active distribution network is to efficiently link power generation with consumer demands, allowing both to decide how best to operate in real-time. Power flow assessment, voltage control and protection require cost-competitive technologies and new communication systems with information and communication technology (ICT) playing a key role.

The realization of active distribution networks requires the implementation of radically new system concepts. **Microgrids** [5–11], also characterized as the "building blocks of smart grids", are perhaps the most promising, novel network structure. The organization of microgrids is based on the **control** capabilities over the network operation offered by the increasing penetration of distributed generators including microgenerators, such as micro-turbines, fuel cells and photovoltaic (PV) arrays, together with storage devices, such as flywheels, energy capacitors and batteries and controllable (flexible) loads (e.g. electric vehicles [12]), at the distribution level. These control capabilities allow distribution networks, mostly interconnected to the upstream distribution network, to also operate when isolated from the main grid, in case of faults or other external disturbances or disasters, thus increasing the quality of supply. Overall, the implementation of control is the key feature that distinguishes microgrids from distribution networks with distributed generation.

From the customer's point of view, microgrids provide both thermal and electricity needs, and, in addition, enhance local reliability, reduce emissions, improve power quality by supporting voltage and reducing voltage dips, and potentially lower costs of energy supply. From the grid operator's point of view, a microgrid can be regarded as a controlled entity within the power system that can be operated as a single aggregated load or generator and, given attractive remuneration, also as a small source of power or ancillary services supporting the network. Thus, a microgrid is essentially an aggregation concept with participation of both supply-side and demand-side resources in distribution grids. Based on the synergy of local load and local microsource generation, a microgrid could provide a large variety of economic, technical, environmental and social benefits to different stake-holders. In comparison with peer microsource aggregation methods, a microgrid offers maximum flexibility in terms of ownership constitution, allows for global optimization of power system efficiency and appears as the best solution for motivating end-consumers via a common interest platform.

Key economic potential for installing microgeneration at customer premises lies in the opportunity to locally utilize the waste heat from conversion of primary fuel to electricity. There has been significant progress in developing small, kW-scale, combined heat and power (CHP) applications. These systems have been expected to play a very significant role in the microgrids of colder climate countries. On the other hand, PV systems are anticipated to become increasingly popular in countries with sunnier climates. The application of micro-CHP and PV potentially increases the overall efficiency of utilizing primary energy sources and consequently provides substantial environmental gains regarding carbon emissions, which is another critically important benefit in view of the world's efforts to combat climate change.

From the utility point of view, application of microsources can potentially reduce the demand for distribution and transmission facilities. Clearly, distributed generation located close to loads can reduce power flows in transmission and distribution circuits with two important effects: loss-reduction and the ability to potentially substitute for network assets. Furthermore, the presence of generation close to demand could increase service quality seen by end customers. Microgrids can provide network support in times of stress by relieving congestion and aiding restoration after faults.

In the following sections, the microgrid concept is clarified and a clear distinction from the virtual power plant concept is made. Then, the possible internal and external market models and regulation settings for microgrids are discussed. A brief review of control strategies for microgrids is given and a roadmap for microgrid development is provided.

1.2 The Microgrid Concept as a Means to Integrate Distributed Generation

During the past decades, the deployment of distributed generation (DG) has been growing steadily. DGs are connected typically at distribution networks, mainly at medium voltage (MV) and high voltage (HV) level, and these have been designed under the paradigm that consumer loads are passive and power flows only from the substations to the consumers and not in the opposite direction. For this reason, many studies on the interconnection of DGs within distribution networks have been carried out, ranging from control and protection to voltage stability and power quality.

Different microgeneration technologies, such as micro-turbines (MT), photovoltaics (PV), fuel cells (FC) and wind turbines (WT) with a rated power ranging up to 100 kW can be directly connected to the LV networks. These units, typically located at users' sites, have emerged as a promising option to meet growing customer needs for electric power with an emphasis on reliability and power quality, providing different economic, environmental and technical benefits. Clearly, a change of interconnection philosophy is needed to achieve optimal integration of such units.

Most importantly, it has to be recognized that with increased levels of microgeneration penetration, the LV distribution network can no longer be considered as a passive appendage to the transmission network. On the contrary, the impact of microsources on power balance and grid frequency may become much more significant over the years.

Therefore, a control and management architecture is required in order to facilitate full integration of microgeneration and active load management into the system. One promising way to realize the emerging potential of microgeneration is to take a systematic approach that views generation and associated loads as a subsystem or a microgrid.

In a typical microgrid setting, the control and management system is expected to bring about a variety of potential benefits at all voltage levels of the distribution network. In order to achieve this goal, different hierarchical control strategies need to be adopted at different network levels.

The possibility of managing several microgrids, DG units directly connected to the MV network and MV controllable loads introduces the concept of multi-microgrids. The hierarchical control structure of such a system calls for an intermediate control level, which will optimize the multi-microgrid system operation, assuming an operation under a real market environment. The concept of multi-microgrids is further developed in Chapter 5.

The potential impact of such a system on the distribution network may lead to different regulatory approaches and remuneration schemes, that could create incentive mechanisms for distribution system operators (DSOs), microgeneration owners and loads to adopt the multi-microgrid concept. This is further discussed in Chapter 7.

1.3   Clarification of the Microgrid Concept

1.3.1   What is a Microgrid?

In scope of this book, the definition from the EU research projects [7,8] is used:

> Microgrids comprise LV distribution systems with distributed energy resources (DER) (micro-turbines, fuel cells, PV, etc.) together with storage devices (flywheels, energy capacitors and batteries) and flexible loads. Such systems can be operated in a non-autonomous way, if interconnected to the grid, or in an autonomous way, if disconnected from the main grid. The operation of microsources in the network can provide distinct benefits to the overall system performance, if managed and coordinated efficiently.

There are three major messages delivered from this definition:

1. Microgrid is an integration platform for supply-side (microgeneration), storage units and demand resources (controllable loads) located in a local distribution grid.
   * In the microgrid concept, there is a focus on local supply of electricity to nearby loads, thus aggregator models that disregard physical locations of generators and loads (such as virtual power plants with cross-regional setups) are not microgrids.
   * A microgrid is typically located at the LV level with total installed microgeneration capacity below the MW range, although there can be exceptions: parts of the MV network can belong to a microgrid for interconnection purposes.
2. A microgrid should be capable of handling both normal state (grid-connected) and emergency state (islanded) operation.
   * The majority of future microgrids will be operated for most of the time under grid-connection – except for those built on physical islands – thus, the main benefits of the microgrid concept will arise from grid-connected (i.e. "normal") operating states.
   * In order to achieve long-term islanded operation, a microgrid has to satisfy high requirements on storage size and capacity ratings of microgenerators to continuous supply of all loads or it has to rely on significant demand flexibility. In the latter case, reliability benefits can be quantified from partial islanding of important loads.
3. The difference between a microgrid and a passive grid penetrated by microsources lies mainly in terms of management and coordination of available resources.
   * A microgrid operator is more than an aggregator of small generators, or a network service provider, or a load controller, or an emission regulator – it performs all these functionalities and serves multiple economic, technical and environmental aims.

- One major advantage of the microgrid concept over other "smart" solutions lies in its capability of handling conflicting interests of different stakeholders, so as to arrive at a globally optimal operation decision for all players involved.

A microgrid appears at a large variety of scales: it can be defined at the level of a LV grid, a LV feeder or a LV house – examples are given in Figure 1.1. As a microgrid grows in scale, it will likely be equipped with more balancing capacities and feature better controllability to reduce the intermittencies of load and RES. In general, the maximum capacity of a microgrid (in terms of peak load demand) is limited to few MW (at least at the European scale, other

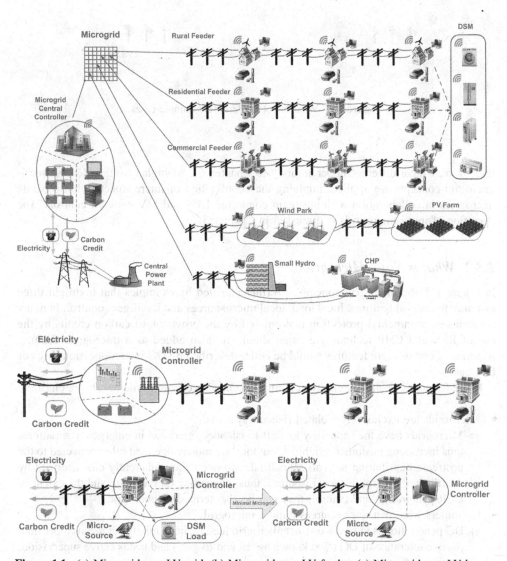

Figure 1.1 (a) Microgrid as a LV grid; (b) Microgrid as a LV feeder; (c) Microgrid as a LV house

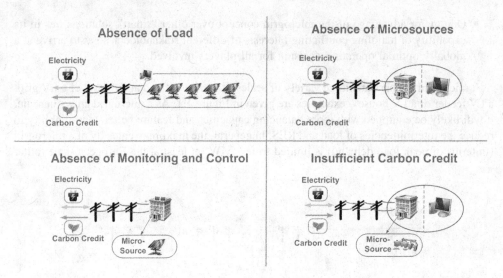

Figure 1.2 What is not a microgrid? Sample cases

regions may have different upper limits, see Chapter 6). At higher voltage levels, multi-microgrid concepts are applied, implying the coordination of interconnected, but separate microgrids in collaboration with upstream connected DGs and MV network controls. The operation of multi-microgrids is discussed in Chapter 5.

1.3.2 What is Not a Microgrid?

In Figure 1.2, the microgrid concept is further clarified by examples that highlight three essential microgrid features: local load, local microsources and intelligent control. In many countries environmental protection is promoted by the provision of carbon credits by the use of RES and CHP technologies; this should be also added as a microgrids feature. Absence of one or more features would be better described by DG interconnection cases or DSI cases.

In the following section, some typical misconceptions regarding microgrids are clarified:

- Microgrids are exclusively isolated (island) systems.
  ⇒ Microgrids have the capability to shift to islanded operation in emergency situations, thus increasing customer reliability, but they are mostly operated interconnected to the upstream distribution network. Small island systems are inherently characterized by coordinated control of their resources, thus, depending on their size and the extent of DER penetration and control, they can be also termed as microgrids.
- Customers who own microsources build a microgrid.
  ⇒ DG penetration is indeed a distinct microgrid feature, but a microgrid means more than passive tolerance of DG (also known as "fit and forget") and needs active supervision, control and optimization.

- Microgrids are composed of intermittent renewable energy resources, so they must be unreliable and easily subject to failures and total black-outs.
  ⇒ A microgrid can offset RES fluctuation by its own storage units (when islanded) or external generation reserves (when grid-connected). Moreover, the microgrid's capability of transferring from grid-connected to island mode actually improves security of supply.
- Microgrids are expensive to build, so the concept will be limited to field tests or only to remote locations.
  ⇒ DER penetration is increasing worldwide. Financial support schemes for RES and CHP have already ensured the basic profitability of such distributed resources; future cost reductions of microgeneration and storage can make microgrids commercially competitive. In any case, the additional cost for transforming a distribution line with DER into a microgrid involves only the relevant control and communication costs. These are easily compensated by the economic advantages of coordinated DER management.
- The microgrid concept is just another energy retailer advertising scheme to increase his income.
  ⇒ Even if an end-consumer chooses not to install photovoltaic panels on his rooftop or hold a share in the community-owned CHP plant, he can still benefit from having more choice of energy supply and of sharing carbon-reduction credits in his bill.
- The microgrid controllers will force consumers to shift their demand, depending on the availability of renewable generation, e.g. to switch on the washing machine at home only when the sun is shining or the wind is blowing.
  ⇒ Demand side integration (DSI) programs in normal commercial and household applications should apply a "load follow generation" control philosophy only to long-term stand-by appliances (such as refrigerators and air-conditioners) or time-insensitive devices (such as water heaters).
- A microgrid is such a totally new idea, that system operators need to rebuild their entire network.
  ⇒ Although new metering, communication and control devices would need to be installed, conversion of a normal "passive" distribution grid to a microgrid does not actually incur too much infrastructure costs on the network operator side – on the contrary, a microgrid can actually defer investment costs for device replacement.
- Microgrid loads will never face any supply interruptions.
  ⇒ "Smooth" (i.e. no loss of load) transition to island operation is only possible with large storage or generation redundancy within a microgrid, thus an islanded microgrid will very probably have to shed non-critical loads according to the instantaneous amount of available resource.

1.3.3 Microgrids versus Virtual Power Plants

A **virtual power plant** (VPP) is a cluster of DERs which is collectively operated by a central control entity. A VPP can replace a conventional power plant, while providing higher efficiency and more flexibility. Although the microgrid and the VPP appear to be similar concepts, there are a number of distinct differences:

- Locality – In a microgrid, DERs are located within the same local distribution network and they aim to satisfy primarily local demand. In a VPP, DERs are not necessarily located on

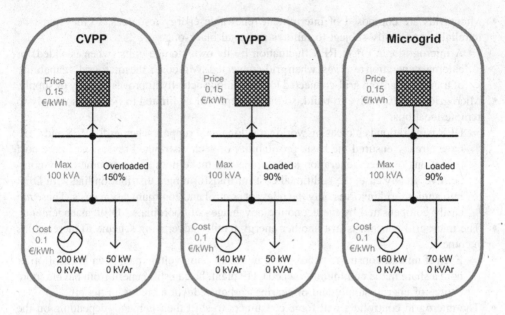

Figure 1.3 Microgrid benefit over commercial and technical VPP due to supply side integration

the same local network and they are coordinated over a wide geographical area. The VPP aggregated production participates in traditional trading in normal energy markets.

- Size – The installed capacity of microgrids is typically relatively small (from few kW to several MW), while a VPP's power rating can be much larger.
- Consumer interest – A microgrid focuses on the satisfaction of local consumption, while a VPP deals with consumption only as a flexible resource that participates in the aggregate power trading via DSI remuneration.

Following on from the definition of a VPP as a commercial entity that aggregates different generation, storage or flexible loads, regardless of their locations, the technical VPP (TVPP) has been proposed, which also takes into account local network constraints. In any case, VPPs, as virtual generators, tend to ignore local consumption, except for DSI, while microgrids acknowledge local power consumption and give end consumers the choice of purchasing local generation or generation from the upstream energy market. This leads to a better controllability of microgrids, as shown in Figure 1.3, where both supply and demand resources of a microgrid can be simultaneously optimized, leading to better DG profitability.

1.4 Operation and Control of Microgrids

1.4.1 Overview of Controllable Elements in a Microgrid

As well as the basic demand and supply resources, a microgrid can potentially be equipped with energy balancing facilities, such as dispatchable loads (e.g. electric vehicles) and substation storage units (Figure 1.4), that could either contribute to minimization of power exchange or maximization of trading profit (in case of free exchange under favorable pricing conditions).

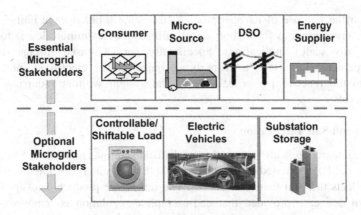

Figure 1.4 Microgrid stakeholders

1.4.1.1 Intermittent RES Units

Controllability of intermittent RES units is limited by the physical nature of the primary energy source. Moreover, limiting RES production is clearly undesirable due to the high investment and low operating costs of these units and their environmental benefits over carbon emission. Consequently, it is generally not advisable to curtail intermittent RES units, unless they cause line overloads or overvoltage problems.

The operation strategy for intermittent RES units can therefore be described as "priority dispatch", that is, intermittent RES units are generally excluded from the unit commitment schedule, as long as they do not violate system constraints. Units with independent reactive power interfaces (decoupled from the active power output) can be included in reactive power dispatch to improve the technical performance of the total microgrid.

1.4.1.2 Dispatchable Microsource and CHP Units

Controllability of CHP units varies according to the way they meet local heat demand – that is, they can be heat-driven, electricity-driven or operated in a hybrid mode. Since most microsource and CHP units are based on rotating machine technology, their reactive power output will be constrained by both the active power output and the apparent power rating.

Due to improved controllability of dispatchable microsource units, a microgrid with multiple microsource units will need to solve the traditional unit commitment problem – albeit at a much smaller scale. On the other hand, the microgrid operator needs to cope with much higher net load variations (i.e. load minus intermittent RES output). At the same time, optimization constraints will likely include grid operating states and emission targets, which add much more complexities to the unit commitment task.

1.4.1.3 Storage Units

Technically, a storage unit could behave either under a load-following paradigm (i.e. balancing applications) or under a price-following paradigm (i.e. arbitrager applications) depending on the purpose of its operation. At the same time, storage units can provide balancing reserves ranging from short-term (milliseconds to minute-level) to long-term (hourly to daily scale) applications. Specifically, for DC-based storage technologies (battery, super-capacitor etc.), a properly designed power electronic interface could contribute to the reactive power balance of the system without incurring significant operational costs.

1.4.1.4 Demand Side Integration

Demand side integration is also referred to as demand side management (DSM) or demand side response (DSR). It is based on the concept [13] that customers are able to choose from a range of products that suit their preferences. The innovative products packaged by energy suppliers will deliver – provided that end-user price regulation is removed – powerful messages to consumers about the value of shifting their electricity consumption. Examples of such offers include

- **time-of-use (ToU):** higher "on-peak" prices during daytime hours and lower "off-peak" prices during the night and at weekends (already offered in some EU member states)
- **dynamic pricing (including real-time pricing):** prices fluctuate to reflect changes in the wholesale prices
- **critical peak prices:** same rate structure as for ToU, but with much higher prices when wholesale electricity prices are high or system reliability is compromised.

The control of customers load can either be

- **manual:** customers are informed about prices, for example on a display, and decide on their own to shift their consumption, perhaps remotely through a mobile phone
- **automated:** customers' consumption is shifted automatically through automated appliances, which can be pre-programmed and can be activated by either technical or price signals (as agreed for instance in the supply contract).

DSI measures in a microgrid are based on forecasts of load and RES outputs and will very probably vary from day to day. A requirement for the successful application of microgrid DSI measures is the adoption of smart metering and smart control of household, commercial and agricultural loads within the microgrid. Depending on the criticality of the target load, DSI measures can generally be divided into shiftable loads and interruptible loads. The integration of DSI measures is expected to maximize their benefits in potential "smart homes", "smart offices" and "smart farms" within microgrids.

1.4.2 Operation Strategies of Microgrids

Currently available DG technologies provide a wide variety of different active and reactive power generation options. The final configuration and operation schemes of a microgrid depend on potentially conflicting interests among different stakeholders involved in electricity supply, such as system/network operators, DG owners, DG operators, energy suppliers, customers and regulatory bodies. Therefore, optimal operation scheduling in microgrids can have economic, technical and environmental objectives (Figure 1.5).

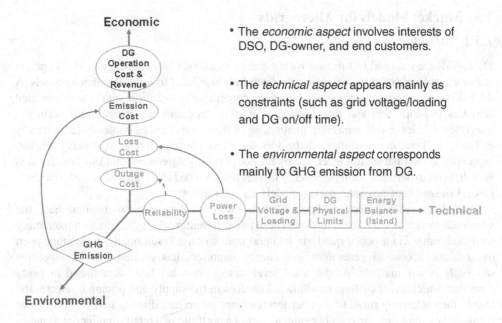

Figure 1.5 Microgrid operation strategies

Depending on the stakeholders involved in the planning or operation process, four different microgrid operational objectives can be identified: economic option, technical option, environmental option and combined objective option.

In the economic option, the objective function is to minimize total costs regardless of network impact/performance. This option may be envisaged by DG owners or operators. DGs are operated without concern for grid or emission obligations. The main limitations come from the physical constraints of DG.

The technical option optimizes network operation (minimizing power losses, voltage variation and device loading), without consideration of DG production costs and revenues. This option might be preferred by system operators.

The environmental option dispatches DG units with lower specific emission levels with higher priority, disregarding financial or technical aspects. This is preferred for meeting environmental targets, currently mainly supported by regulatory schemes. DG dispatch is solely determined by emission quota; only DG physical limitations are considered.

The combined objective option solves a multi-objective DG optimal dispatch problem, taking into account all economic, technical and environmental factors. It converts technical and environmental criteria into economic equivalents, considering constraints from both network and DG physical limits. This approach could be relevant, for instance, to actors that participate not only in classical energy markets, but also in other potential markets for provision of network services and emission certificates.

1.5 Market Models for Microgrids

1.5.1 Introduction

Microgrids are required to function within energy markets. Current energy markets operate with various levels of complexity ranging from fully regulated to fully liberalized models. A related issue is the interdependency between competitive (generation, retail) and regulated activities (transmission and distribution) in market structures that range from vertically integrated utilities to full ownership unbundling of these activities, as implemented currently in Europe. Therefore, definition of the various actors participating in the energy market, especially at the distribution level, are by far not generally agreed. Within this context, it is very difficult to define a "generic" energy market model. The following discussion provides a general overview of possible market models for microgrids.

Depending on the operational model, two major markets can be distinguished: the wholesale market and the retail market. These two different markets can function interacting with each other via a pool or/and via bilateral transactions. Traditionally, mandatory open transmission access to generators and energy importers has created more competitive wholesale power markets. At the retail level, competition has been established in many countries, which gives customers additional choices in the supply and pricing of electricity. Due to their relatively small sizes, microgrids cannot participate directly in the wholesale or retail market, thus they can possibly enter as part of a portfolio of a retail supplier or an energy service company (ESCO). Direct control of DERs by DSOs is another possible model. Fair competition in the retail sector is expected to serve as the basis of microgrid adoption, which assumes a suitable regulatory environment and an ICT infrastructure capable of supporting the market operation (e.g. electronic/smart meters and a standardized interaction between retail companies and DSOs), as further discussed in Chapter 2. These are essential conditions for the successful implementation of microgrids.

The relevant key actors in the energy market are as follows:

- **Consumer:** The consumer can represent a household or a medium or small enterprise. Typically the consumer has a contract with a retail company for his energy supply. Furthermore, the consumers, or anyone connected to the distribution network, need to pay fees to the distribution network owner for using the network.
- **DG owner/operator:** Typically the owners of the DG units are also responsible for their operation. DG owners inject their production to the network, possibly enjoying priority dispatch and fixed feed-in tariffs, especially for RES-based DG or they might have contracts with a retail company. They may pay distribution network charges. It is assumed that some or all of the DG units will be equipped with at least some monitoring and possibly control capabilities.
- **Prosumer:** This is the special case of consumers who have installed small DG in their premises. The local DG production can cover, wholly or partly, the consumption of the owner, and the surplus can be exported to the main power grid. Alternatively, the total local production can also be sold directly to the main power grid enjoying favorable feed-in tariffs.
- **Customer:** Consumers, DG owners/operators and prosumers are termed customers.
- **Market regulator:** A regulatory authority for energy is an independent body responsible for the open, fair and transparent operation of the market, ensuring open access to the

network and efficient allocation of network costs. Depending on local conditions, it also approves the level of network usage charges and in some cases end-user prices.

- **Retail supplier, energy service company (ESCO):** A supplier directly interacts with the customers and has contracts with them. Its main duties are to provide electricity and possibly other energy services to its customers. Regarding electricity retail activities, the supplier acquires energy from different sources, such as the wholesale market or spot market or local DG production. Unless partly or wholly regulated, the supplier determines the energy prices of the electricity delivered to the customer, which may vary depending on time and location. In the context of microgrids, suppliers are the actors to maximize the value of the aggregated DER participation in local energy markets, thus maximizing the microgrid's value. Suppliers are also responsible for translating the complexity and sophistication of the retail market, including the demand response market and DER remuneration schemes, into simple forms that customers demand, by "packaging" attractive products. The products offered should be easy for the customer to understand and suppliers should effectively manage any complexity in costs (e.g. variable grid tariffs). Depending on the regulatory framework and the market conditions, suppliers might offer, next to the electricity supply, a broader range of energy services featuring better reliability, appealing (green) energy production and convenience. In the future, microgrids can be part of an ESCO commercial business portfolio, which includes a broad range of comprehensive energy solutions, aiming to reduce the holistic energy cost of a group of buildings or a building complex. Depending on the business case, the supplier/ESCO might need to install the necessary home gateway that will be responsible for appliance management.
- **Distribution system operator (DSO):** The distribution system operator is the actor responsible for the operation, maintenance and development of the distribution network in a given area. Usually the DSO (a) manages the HV, MV and LV distribution systems, (b) is obliged to deliver electricity to consumers or absorb energy from DG/RES and (c) being a regulated entity, it is not involved in any retail activity. In a future power system characterized by smart grids and load flexibility, DSOs will provide the playing field where suppliers can offer innovative products, and they will therefore play the role of neutral market facilitators. Where applicable, the role of DSOs as metering operators is crucial for the efficient delivery of energy services offered by suppliers on a competitive basis. Suppliers will need timely, transparent and non-discriminatory access to commercial data, while DSOs have access to the technical data necessary to manage their grid effectively. Taking into account that microgrids provide considerable flexibility, DSOs will contract both with suppliers managing microgrid customers and with individual distributed generators to utilize their flexibility for local network balancing. Where applicable, network tariffs should support microgrids' commercial participation and more generally, the development of demand response products. The direct control of demand response or DER (e.g. storage, installed at the distribution level) without involving suppliers is another possibility, although it would threaten the suppliers' ability to balance their portfolios, increasing the need for contracting reserves, and this in turn would lead to increased costs for consumers. In the future, it is anticipated that a new set of agreements between suppliers and DSOs will ensure that customers benefit from proper functioning of the market, smooth processes and a secure and reliable electricity supply; suppliers will market new products and optimize their supply and balancing portfolio, while DSOs can guarantee local grid stability and security of supply through system services.

- **Microgrid operator:** The microgrid operator is responsible for the operation, maintenance and development of the local (LV) distribution network forming the microgrid. Depending on the business model, discussed in Section 1.4.2, this role can be assumed by the local DSO or it can be a dedicated, independent DSO who acts locally on behalf of the microgrid customers.

Microgrids' participation in energy markets can be analyzed from the generation and the demand side.

1.5.1.1 DG/RES Generation

As discussed in the previous section, the production of DG sources within a microgrid can be dispatchable or intermittent for certain RES technologies, such as PVs and small wind turbines. Today, in most countries, there is no requirement to control the production of RES units at the distribution level. The DSO is obliged to absorb all the energy produced, except for security reasons, and most RES technologies enjoy favorable feed-in-tariffs for their production. This framework provides no incentives to the DG owners for active management of their units or for actually integrating them in microgrid operation.

1.5.1.2 Demand Side

Demand side integration is an important feature of microgrid operation, as discussed in Section 1.4.1. Demand response can be classified according to the way load changes are induced [14]:

- **Price-based demand response** implementations induce changes in usage by customers in response to time-based changes in the prices they pay. This includes real-time pricing, critical-peak pricing and time-of-use rates.
- **Incentive-based demand response** programs are facilitated by utilities, retail companies or a regional DSO. These programs give customers load reduction incentives that are separate from, or in addition to, their retail electricity rates, regardless of whether the rates are fixed (based on average costs) or time-varying. Load reductions are coordinated as requested by the program operator and can include **direct load control**, that is, a program through which the program operator remotely shuts down or cycles a customer's electrical equipment (e.g. air conditioner, water heater, space heating) at short notice.

The goal of both policies is to smooth the daily load curve of the system by shifting load. This includes peak shaving (limit consumption during hours of peak load) and valley filling (increase consumption during the hours of low demand). Load shifting can be also used to compensate the volatility in RES production or in the electricity prices on the spot market. In general, flexible demand is an obvious source of "negative" power reserves, offering a great potential that has not so far been adequately explored in practice.

1.5.1.3 Ancillary Services

"Ancillary services are all services required by the transmission or distribution system operator to enable them to maintain the integrity and stability of the transmission or distribution system as well as the power quality" [15]. The services include both mandatory services and services subject to competition. The increased penetration of DER in the distribution network combined with the objective of more reliable and cost-efficient network operation, have opened up new opportunities for meeting system obligations and providing grid services (reactive power, voltage support, congestion management) which are required by DSOs at the distribution level. Ancillary services are classified according to the operation modes of microgrids (normal operation or emergency operation/grid-connected operation or islanded operation), as follows:

- grid-connected operation:
  - frequency control support
  - voltage control support
  - congestion management
  - reduction of grid losses
  - improvement of power quality (voltage dips, flicker, compensation of harmonics)
- islanded operation:
  - black start
  - grid-forming operation:
    - frequency control
    - voltage control

The participation of microgrids in local ancillary service markets can be seen as a great opportunity, especially regarding active reserves and voltage support. DER capabilities and detailed technical solutions for the provision of ancillary services are discussed in the following chapters of this book, but the necessary regulatory environment needs to be applied, and a number of critical questions need to be answered, such as:

- What is the cost of provision of ancillary services?
- Who is responsible for coordinating the associated functions?

1.5.2 Internal Markets and Business Models for Microgrids

The internal market of a microgrid mainly relates to the ownership and business models established between major stakeholders involved in the operation of a microgrid, such as local consumers, microsources, DSO and energy supplier. A microgrid's internal market mainly determines the amount and direction of cash flows within the microgrid. It also defines which entity will participate in external markets, as representative of the whole group of stakeholders. The structure of an internal market, however, does not necessarily impact the choice of microgrid operation strategy or its collective behavior, as seen from the external grid; thus two microgrids with completely different ownership properties might behave very similarly to the external energy market.

Microgrid structures can be differentiated by the level of their DER aggregation: they range from a simple collection of independent market players to a collaboration that encompasses all demand- and supply-side entities. Similar to the unbundling of central generation from transmission network, the operation structure of a microgrid will be mainly decided by the ownership of microsources, that is, by the DSO, the end consumer, the energy supplier or directly by the microsource operator as independent power producer (IPP) (Figure 1.6).

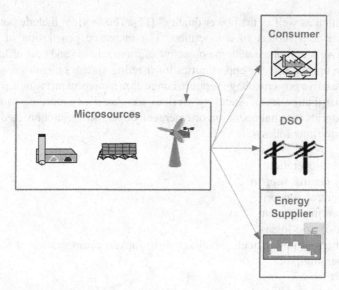

Figure 1.6 Sample micro-source ownership possibilities in a microgrid

Although theoretically there could be numerous forms of microgrids, three typical setups can be identified, named as the DSO monopoly, free market and prosumer consortium models. These are discussed in the following subsections. In all models it is assumed that all the required metering, monitoring and control functionalities are provided.

1.5.2.1 DSO Monopoly Model

In a DSO monopoly microgrid, the DSO is part of a vertically integrated utility, so it not only owns and operates the distribution grid, but also fulfills the retailer function of selling electricity to end consumers. Within this single-player context, integration and operation of microsources is most conveniently also undertaken by the DSO, which leaves almost all technical and financial consequences (i.e. both costs and benefits) of microgrid conversion to DSO responsibility (Figure 1.7). In unbundled systems, DSOs might not be allowed to own and operate a DER, but a DER used exclusively for supporting distribution system operation might be an exception.

In a DSO monopoly microgrid, microgeneration units tend to be larger, and storage units tend to be located at central substations.

With the business model of Figure 1.7, a versatile "extended" DSO plays the role of both physical and financial bridge between the overlying grid and end consumers. Microsource control decisions are basically made within the framework of DSO functionalities. There is no room for a local service market except for tariff-driven DSI programs.

In general, a DSO monopoly microgrid is most likely to be built upon a technically challenged distribution grid with aging infrastructure, maintenance and/or supply quality problems. The investment decision in microsource units by a DSO (if allowed by the market regulator) can be generally explained as an alternative to more expensive network solutions,

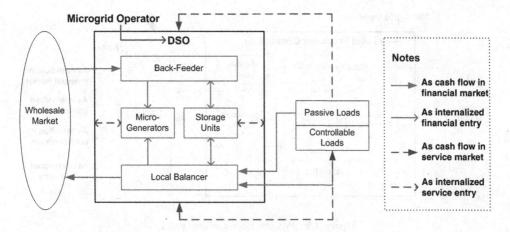

Figure 1.7 DSO monopoly model

such as upgrading overloaded lines to overcome thermal constraints. Therefore, the potential profitability of selling microsource energy to local consumers may not be the primary consideration.

1.5.2.2 Liberalized Market Model

In liberalized markets, microgrids (Figure 1.8) can be driven by various motives (economic, technical, environmental etc.) from various stakeholders (suppliers, DSO, consumers etc.). As discussed in Section 1.5.1, suppliers/ESCOs are the actors best suited to maximize the value of the aggregated DER participation in local liberalized energy markets, that is, for the commercial participation of microgrids. The daily operational decisions depend on real-time negotiations (i.e. interest arbitration) of all parties involved. In this case, a microgrid operator or central controller (MGCC) is responsible for microgrid local balance, import and

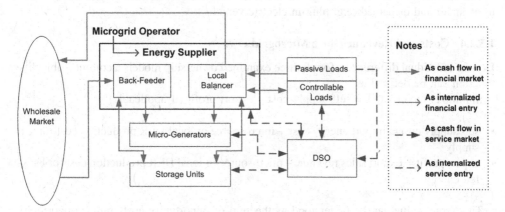

Figure 1.8 Free market model

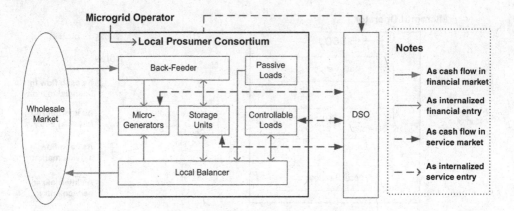

Figure 1.9   Prosumer consortium model

export control, technical performance maintenance and emission level monitoring. In a free market microgrid, DG and storage units can vary in forms, sizes and locations.

1.5.2.3   Prosumer Consortium Model

A prosumer consortium microgrid (Figure 1.9) is most likely to be found in regions with high retail electricity price and/or high microgeneration financial support levels. In this case, single or multiple consumers own and operate microsources to minimize electricity bills or maximize sales revenue from energy export to the upstream network. This type of microgrid tends by nature to minimize the use of the distribution grid (which leads to a reduction in the use of system revenues) and may neglect network constraints (i.e. DG hosting capacity) during its design. DSOs can only passively influence the operation of a prosumer consortium microgrid by imposing requirements and charges on the microsource owners, but will not be able to benefit from the local trading process.

In a prosumer consortium microgrid, microgeneration tends to be smaller, and storage tends to be small and dispersed (e.g. plug-in electric vehicles).

1.5.2.4   Costs and Revenues for a Microgrid Owner

Despite individual differences, for all three examined ownership models a common objective function can be defined as

minimize: target value = self supply cost − export profit (if applicable)

- self supply cost = import energy cost + microsource or local DER production cost for self supply
- export profit = export sales revenue − microsource or local DER production cost for back-fed energy

This target value can be understood as the sum of opportunity costs minus opportunity revenues obtained from optimal real-time dispatch decisions between internal (microgrid) and external (market) resources. Of course, for each microgrid ownership model examined, extra internal cost entries will arise due to the differences of corresponding interest allocation models.

1.5.3 External Market and Regulatory Settings for Microgrids

One critical factor of the financial and technical feasibility of microgrids is external market operation and the associated regulatory rules. As all investment decisions are de facto based on expected profitability, the external environment of a microgrid has a major impact over whether or not a commercial-level microgrid will be realized. Public policy support for microgrids in order to promote technological evolution will also prove to be critical for creating a level playing field for both existing and new players in the market.

As briefly introduced in the previous section, there are two main types of external markets with impact on a microgrid: the energy market and the ancillary services market. It is clear that the ancillary services market can be an ideal playing field for microgrids owing to their controllability, when properly designed. In the following part of this section focus is placed on the three main factors of external energy market operation that have critical influence on a microgrid:

1. satisfaction of local consumption
2. recognition of locational value
3. time of use tariffs for local energy

1.5.3.1 Satisfaction of Local Consumption

Satisfaction of local consumption implies that the power produced by microgeneration is used to supply partly or wholly in-site consumption. In such a case, there is no requirement for separate metering of microsource generation (also called "net-metering"). However, on-site generation and on-site load need to be metered separately when microsource units appear as independent generators that sell all their production directly to the network and are not financially related to end consumers. In this case, local consumption is a market opportunity that can be easily overlooked by all players (Figure 1.10).

There are two main advantages of promoting local consumption satisfaction within a microgrid:

1. End consumers are provided with more choices in retail power supply.
2. Microsource operators have the possibility to obtain quasi-retail prices via selling locally to minimize network charges.

The local retail market concept is therefore directly linked to the local consumption mechanism, which can also be seen as a two-sided hedging tool for both demand and supply players for reducing market risk: consumers can use the local market to hedge against high market price, while microsources can use the local market to hedge against low market price.

1.5.3.2 Recognition of Locational Value

The microgrid operator or retail company that undertakes the microgrid market participation is likely to trade with the external market on a daily basis, which implies both purchase and

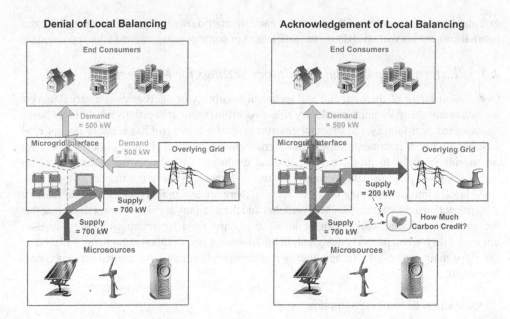

Figure 1.10 Impact of local balancing

selling of electricity. Under this general setting, a microgrid is likely to be faced with wholesale prices for export and (quasi) retail prices for import (**directional pricing**), unless a different regulatory framework is in place.

In the microgrid context, the difference between purchase and selling prices is attributed to the network charges or use of system (UoS) charges, which is applied on top of the electricity prices in the wholesale market. However, because microsources are located closer to the consumption, they do not require the same transmission and distribution network infrastructures. The locational value of the microsources could be recognized by allowing a total or partial exemption of UoS charge that can eventually improve the profitability of these microsources against the more competitive large, central units. In fact, the exemption from UoS charges can be seen as a security of supply compensation to be paid by the DSOs, in order to avoid or defer investments for expensive network upgrades.

Depending on the level of political and social support for microgrids, DSOs may also agree to lower UoS charges imposed on power trades between microgrids and external generators or loads – especially when the power is exchanged between one microgrid and another, or with a DG or controllable load at the MV level. The basic reasoning behind this is to extend the internal production–consumption balance of a microgrid to multi-microgrids – the aggregation of microgrids in conjunction with an MV level DG and controllable loads that are comparable in size with a complete LV microgrid. In this way (shown in Figure 1.11), selling and buying prices both within and outside of a microgrid can be unified to promote bidirectional trading on a real-time basis (**uniform pricing**).

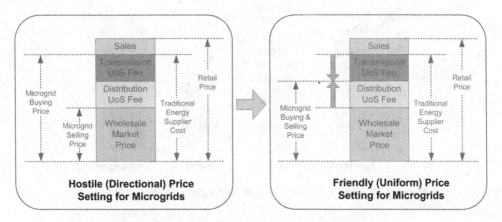

Figure 1.11 Energy pricing schemes for microgrids

1.5.3.3 Fixed versus Time of Use Tariffs

Electricity pricing can be fixed, that is, a microgrid is subject to a fixed selling price and a fixed buying price (under a uniform pricing scheme both prices coincide) for participation in the energy market or varying with time – time of use (ToU) prices – that is, a microgrid obtains hourly (or time-of-day block) prices for both selling and buying electricity. It is clear that the adoption of ToU prices offers greater flexibility and favors local resources and thus the adoption of microgrids.

1.5.3.4 Example

The following practical example demonstrates in a simple way the importance of the above factors on the profitability of microgrids. By combining the satisfaction of local consumption, the recognition of locational value and the adoption of ToU tariff criteria, eight potential external market environments can be derived. Out of them one reference case and three main microgrid-enabling cases are shown in Table 1.1.

The reference case refers to a distribution network where local DG production is ignored and local consumption is supplied by energy imported at fixed retail prices. In the (forced) island case, the local DG units have economic incentive to feed local consumption at their production costs, but are discouraged from exporting due to low remuneration tariff. In the hybrid case, energy from the external market, if cheaper, is bought at real-time retail price,

Table 1.1 Microgrid pricing schemes.

Case	Local Consumption	Locational Value	Tariffs wrt. Time
Reference	No	No	Fixed
Island	Yes	No	Fixed
Hybrid	Yes	No	ToU
Exchange	Yes	Yes	ToU

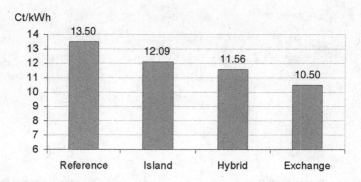

Figure 1.12 Per kWh electricity costs under different pricing strategies

while in the exchange case the consumption is supplied by the optimal combination of energies bought from the internal and external market at "uniform" prices, assumed to be the average between wholesale and retail prices. For both hybrid and exchange cases, power export becomes a feasible option due to the potential price peaks. In order to illustrate potential economic consequences of the four proposed pricing scenarios, a four-hour, two-unit dispatch problem is analyzed with the following assumptions: mean wholesale price is assumed to be 7.5 Ct/kWh, mean retail price is 13.5 Ct/kWh. Unit 1 is rated at 30 kW with 12 Ct/kWh, unit 2 is rated at 15 kW with 15 Ct/kWh. Hourly demands are assumed to be 15 kWh, 35 kWh, 25 kWh and 5 kWh from hours 1 to 4.

In Figure 1.12, the per kWh electricity costs in the sample microgrid are compared for the four scenarios. It is shown that the cost decreases as price flexibility and transparency improves. The exchange case is proven to provide the best financial conditions for a microgrid, as it provides the lower cost for electricity supply within the system. The (forced) island case provides the worst financial conditions for the formation of the microgrid, while in the hybrid case the cost lies between the island and exchange cases following variations of real-time prices.

This example suggests that a large installation of microsource capacity that exceeds peak load demand could provide export potential of excess generation. If the microsource units in a microgrid are too small to achieve reversal of power flow, the eventual cost difference between hybrid and exchange case will be significantly reduced and might even lead to lower costs of the hybrid case compared to the exchange case.

1.6 Status Quo and Outlook of Microgrid Applications

Currently, an increasing number of microgrid pilot sites can be observed in many parts of the world, as described in Chapter 6 of this book. It is true, however, that up to now, **cost**, **policy** and **technology** barriers have largely restrained the wide deployment of microgrids in distribution networks owing to their limited commercial appeal or social recognition. However, these three barriers are currently undergoing considerable changes – they are very likely to turn into key enablers in the future, eventually leading to a widespread microgrid adoption worldwide.

Firstly, the cost factor might prove to be the most effective driving force for microgrids in the very near future. This might happen not only because of the reduction of microsource costs, but also because of the relative changes of external opportunity costs due to economic (fluctuating market prices), technical (aging of network infrastructure) and environmental (emission trading) factors.

When microsource penetration at a LV grid becomes significant, participants in the electricity retail business will consider the aggregated power from small generators as a new market opportunity. Unlike in the case of VPP, microgrid stakeholders will eventually recognize a unique feature of aggregated microsource units, namely locality: the microsource units can potentially sell directly to end consumers in an "over-the-grid" manner. In order to turn this potential into reality, however, the second factor – appropriate policy and regulatory environment – is needed to enable the operation of a local market within a microgrid.

Finally, the adoption of favorable selling prices in local retail markets will attract even more microsource units, allowing the microgrid to operate islanded, if beneficial. With the help of

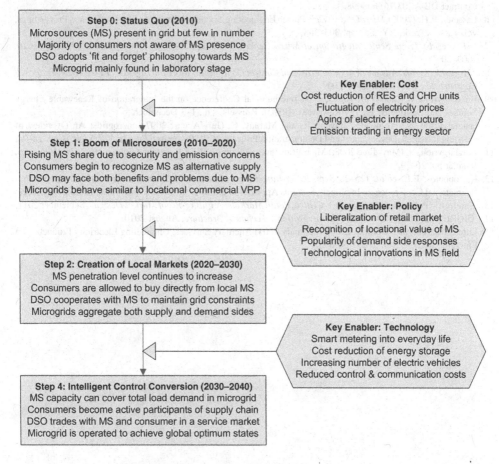

Figure 1.13 Roadmap for microgrid development

smart metering, control and communication technologies, the microgrid operator will eventually be able to coordinate a large consortium of intermittent and controllable micro-source units, as well as central and distributed storage devices, to achieve multiple objectives and, at the same time, to cater for the interests of different stakeholders. Figure 1.13 provides an exemplary roadmap for microgrid development in Europe [8].

References

1. CIGRE (June 2011) Report on behalf of the Technical Committee. "Network of the Future", Electricity Supply Systems of the Future, *Electra*, No 256.
2. *Smartgrids, European Technology Platform on* Vision and Strategy for Europe's Electricity Networks of the Future. s.l.: EUR 22040, 2006.
3. Peças Lopes, J.A., Hatziargyriou, N., Mutale, J. *et al.* (2007) Integrated distributed generation into electric power systems: A review of drivers, challenges and opportunities. *Elsevier Electr. Pow. Syst. Res.*, **77** (9).
4. Pepermans, G., Driesen, J., Haeseldonckx, D. *et al.* (2005) Distributed generation: definition, benefits and issues. *Int. J. Energy Policy*, **33** (6), 787–798.
5. Lasseter, R., Akhil, A., Marnay, C. *et al.* (April 2002) *White paper on Integration of Distributed Energy Resources – The CERTS MicroGrid Concept.* s.l.: Office of Power Technologies of the US Department of Energy, Contract DE-AC03-76SF00098.
6. Lasseter, R.H. (2002) *MicroGrids.* IEEE Power Engineering Society Winter Meeting Conference Proceedings, vol. 1., New York, NY: s.n., pp. 305–308.
7. *"Microgrids: Large Scale Integration of Micro-Generation to Low Voltage Grids", ENK5-CT-2002-00610.* 2003–2005.
8. *"More microgrids: Advanced Architectures and Control Concepts for More microgrids", FP6 STREP, Proposal/ Contract no.: PL019864.* 2006–2009.
9. Hatziargyriou, N. (2004) *MicroGrids.* 1st International Conference on the Integration of Renewable Energy Sources and Distributed Energy Resources (IRED). Brussels: s.n., 1–3 December.
10. Hatziargyriou, N., Asano, H., Iravani, R. and Marnay, C. (July/August 2007) microgrids: An Overview of Ongoing Research, Development and Demonstration Projects. *IEEE Power and Energy, Vol.* 5, *Nr.* 4, pp. 78–94.
11. Hatziargyriou, N. (May–June 2008,) Microgrids [guest editorial]. *IEEE Power and Energy Magazine, Volume* 6, *Issue* 3, pp. 26–29.
12. Karfopoulos, E.L. *et al.* (25–28 Sept 2011) *Introducing electric vehicles in the microgrids concept.* 16th International Conference on Intelligent Systems Application to Power Systems, Hersonissos, Heraklion, Greece.
13. Eurelectric Policy Paper. *"Customer-Centric Retail Markets: A Future-Proof Market Design".* September 2011.
14. CIGRE WG C6.09. *"Demand Side Integration", Technical Brochure*, August 2010.
15. Eurelectric, Thermal Working Group (February 2004) Ancillary Services: Unbundling Electricity Products – an Emerging Market, Ref: 2003-150-0007.

2

Microgrids Control Issues

Aris Dimeas, Antonis Tsikalakis, George Kariniotakis and George Korres

2.1 Introduction

The notion of control is central in microgrids. In fact, what distinguishes a microgrid from a distribution system with distributed energy resources is exactly the capability of their control, so that they appear to the upstream network as a controlled, coordinated unit [1,2]. Primary control of DER is discussed in Chapter 3. This chapter focuses on secondary control or energy management issues [3]. Effective energy management within microgrids is a key to achieving vital efficiency benefits by optimizing production and consumption of heat, gas, and electricity. It should be kept in mind, that the microgrid is called to operate within an energy market environment probably coordinated by an energy service provider/company (ESCO), who will act as an aggregator of various distributed sources and probably a number of microgrids.

The coordinated control of a large number of DERs can be achieved by various techniques, ranging from a basically centralized control approach to a fully decentralized approach, depending on the share of responsibilities assumed by a central controller and the local controllers of the distributed generators and flexible loads. In particular, control with limited communication and computing facilities is a challenging problem favoring the adoption of decentralized techniques. Complexity is increased by the large number of distributed resources and the possible conflicting requirements of their owners. The scope of this chapter is to present an overview of the technical solutions regarding the implementation of control functionalities. Thus, not only the electrical operation will be presented but also issues regarding the ICT challenges.

2.2 Control Functions

Before analyzing the technical implementation of control, a general overview of the main control functionalities in a microgrid is presented. These functionalities can be distinguished

Microgrids: Architectures and Control, First Edition. Edited by Nikos Hatziargyriou.
© 2014 John Wiley & Sons, Ltd. Published 2014 by John Wiley & Sons, Ltd.
Companion Website: www.wiley.com/go/hatziargyriou_microgrids

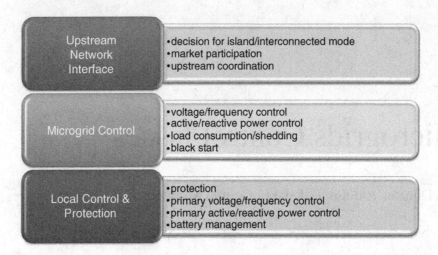

Upstream Network Interface	•decision for island/interconnected mode •market participation •upstream coordination
Microgrid Control	•voltage/frequency control •active/reactive power control •load consumption/shedding •black start
Local Control & Protection	•protection •primary voltage/frequency control •primary active/reactive power control •battery management

Figure 2.1 Overall system functionalities

in three groups, as shown in Figure 2.1. The lower level is closely related to the individual components and local control (micro sources, storage, loads and electronic interfaces), the medium level to the overall microgrid control and the upper level to the interface to the upstream network.

More specifically:

Upstream network interface

The core interaction with the upstream network is related to market participation, more specifically the microgrid actions to import or export energy following the decisions of the ESCO. Owing to the relatively small size of a microgrid, the ESCO can manage a larger number of microgrids, in order to maximize its profit and provide ancillary services to the upstream network. The operation of multi-microgrids is discussed in Chapter 5.

Internal microgrid control

This level includes all the functionalities within the microgrid that require the collaboration of more than two actors. Functions within this level are:

- load and RES forecast,
- load shedding/management,
- unit commitment/dispatch,
- secondary voltage/frequency control,
- secondary active/reactive power control,
- security monitoring,
- black start.

Local control

This level includes all the functionalities that are local and performed by a single DG, storage or controllable load, that is:

- protection functions,
- primary voltage/frequency control,

- primary active/reactive power control,
- battery management.

It should be noted that these functionalities are relevant to the normal state of operation. They might need to change in critical or emergency states, as discussed in Chapter 5.

Chapter 2 focuses on internal microgrid control (management) in the normal state of operation. The normal state covers both islanded and interconnected mode and does not deal with transition to island mode. The role of information and communication technology is critical for the relevant control functions.

2.3 The Role of Information and Communication Technology

Information and communication technology (ICT) is a critical component of future power networks. Beyond any doubt, the control and operation of the future power grids, including microgrids, needs to be supported by sophisticated information systems (ISs) and advanced communication networks. Currently, several technologies have been used or tested at distribution systems and it is expected that their usage will become more extensive during the coming years. The usual approach is to use existing solutions as a starting point, in order to develop new applications for microgrids. The main technological areas are as follows:

Microprocessors
Modern microprocessors are utilized extensively within microgrids providing the ability to develop sophisticated inverters and to develop load controllers or other active components within microgrids. An interesting characteristic of the new version of microprocessors is that they provide adequate processing power, communication capabilities and sophisticated software-middleware at low prices.

Communication
The past decade has been characterized by developments in communication networks and systems. These networks provide sufficient bandwidth and can offer several services to the users. It is obvious that active control of microgrids will be based on existing communication infrastructures, in order to reduce the cost.

Software
Service oriented architectures (SOA) is the modern trend in building information systems. The core of this approach is the web service. The W3C (World Wide Web Consortium [4]) defines a "web service" as "a software system designed to support interoperable machine-to-machine interaction over a network". There are many definitions of the concept of SOA. For the purpose of this chapter an SOA is defined as a set of web services properly organized in multiple layers, capable of solving a set of complex problems.

The internet of energy
The internet of energy is the usage of technologies, developed for the world wide web, that allows avoiding installation and maintenance costs of dedicated devices for the control of DGs and loads. With this approach, all the applications for household control could take the form of a piece of software running on a device with processing capabilities: a smart TV or the internet

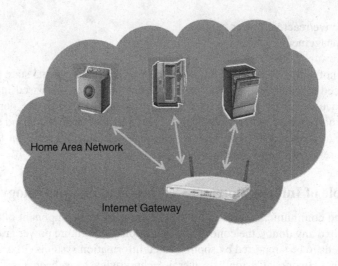

Figure 2.2 The smart home

gateway, for example. It is also assumed that the next generation of home appliance will be equipped with the necessary interfaces to allow remote access via the home area network (Figure 2.2).

A significant part of the necessary technology already exists: internet gateways, IPv6, embedded processors, smart phones, corresponding operating systems and so on. Furthermore, several houses nowadays are quite automated, using advanced sets of home cinemas with internet connection, advanced wireless alarm systems, central automated air-conditioning systems and so on.

2.4   Microgrid Control Architecture

2.4.1 Hierarchical Control Levels

There is no general structure of microgrid control architecture, since the configuration depends on the type of microgrid or the existing infrastructure. Before analyzing the microgrid control and management architecture, let's have a look at today's distribution systems. Figure 2.3 presents the major parts of the control and management infrastructure of a typical distribution system with increased DG penetration. We can distinguish the distribution management system (DMS) and the automated meter reading (AMR) systems. The DMS is mainly responsible for the monitoring of the main HV/MV and maybe some critical MV/LV substations. The hardware system consists of the main server and several remote terminal units (RTUs) or intelligent electronic devices (IEDs) spread across the distribution system. Usually the DMS does not control the DGs/RESs (except for some large installations in certain cases) or the loads. Typical control actions are network reconfiguration, by switching operations in the main feeders, and voltage control via capacitor switching or perhaps transformer tap changing (mostly manually). The AMR system is responsible for the collection of electronic meter readings and is used mainly for billing purposes. In Figure 2.3, we do not consider the existence of the advanced meter infrastructure (AMI), since this is

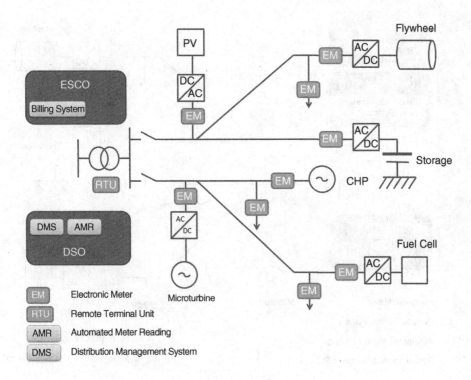

Figure 2.3 Typical distribution system management structure

considered next, as part of the microgrid control system. By AMI, we also mean the capability of controlling some loads locally, either directly via the meter or via the home area network, in which case the electronic meter is the gateway.

As discussed in Section 1.5.1 the DSO is responsible for managing and controlling the distribution system and is also responsible for collecting the energy metering data, although in some countries meter reading can be handled by an independent entity. The DSO sends the metering data to the supplier/ESCO, who is a market player and is responsible among other things for the billing of customers.

The structure shown in Figure 2.3 is not sufficient for microgrid management, since it provides limited control capabilities, especially within a market environment. Thus, it is important to introduce a new control level locally at DG and loads, capable of meeting the goals of

- enabling all relevant actors to advanced market participation
- being scalable in order to allow the integration of a large number of users (scalability)
- allowing integration of the components of different vendors (open architecture)
- ease of installing new components (plug-and-play)
- ease of integrating new functionalities and business cases (expandability).

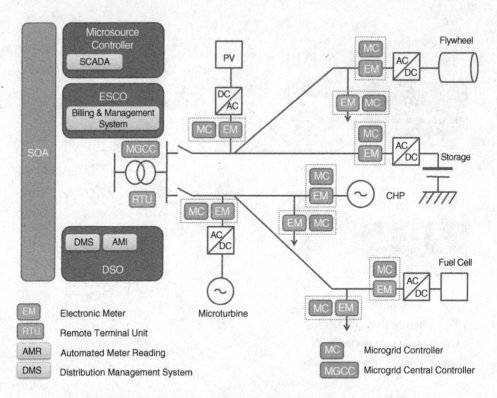

Figure 2.4 Typical microgrid management structure

Using the local control level, a more complicated, hierarchical architecture is introduced in Figure 2.4. This architecture comprises the following:

The **microsource controller (MC)** which is responsible for controlling and monitoring distributed energy resources, such as DGs, storage devices and loads, including electric vehicles. The MC could be a separate hardware device or piece of software installed in either the electronic meter, the DG power electronic interface or any device in the field with sufficient processing capacity. This is shown as a dashed frame that surrounds both the MC and the EM.

The **microgrid central controller (MGCC)** provides the main interface between the microgrid and other actors such as the DSO or the ESCO, and can assume different roles, ranging from the main responsibility for the maximization of the microgrid value to simple coordination of the local MCs. It can provide setpoints for the MCs or simply monitor or supervise their operation. It is housed in the MV/LV substation and comprises a suite of software routines of various functionalities depending on its role.

The **distribution management system (DMS)**, discussed previously, is responsible among others for the collaboration between the DSO, the ESCO and the microgrid operator. The existence of a backbone system, a platform, based on service oriented architecture is assumed for the integration of its functionalities. In some cases, the MGCC software can be integrated in this platform.

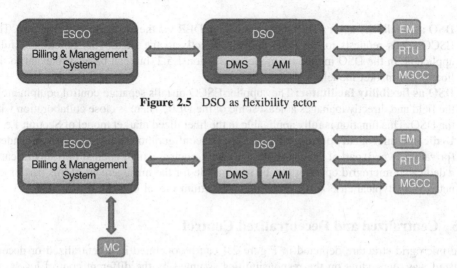

Figure 2.5   DSO as flexibility actor

Figure 2.6   DSO as flexibility facilitator

2.4.2 Microgrid Operators

The microgrid operator, introduced in Section 1.5.1 can be further distinguished depending on the type of microgrid and the roles of the DSO and supplier/ESCO. The role of the DSO as a "flexibility facilitator" or "flexibility actor" is central in these distinctions. Based on Figure 2.4, three main general configurations, as presented in Figures 2.5–2.7, can be identified. It should be noted that the aim of these figures is to show the flow of information among actors and not to present microgrid business models.

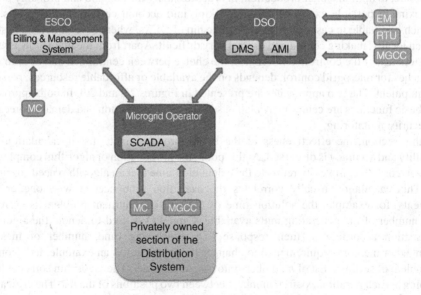

Figure 2.7   Dedicated microgrid operator

- **DSO as flexibility actor:** The DSO controls the DER via the available infrastructure. The ESCO sends requests to the DSO and not directly to the field. This function is fully applicable in the DSO monopoly model of Section 1.5.2, but can be also relevant to the liberalized market model.
- **DSO as flexibility facilitator:** The supplier/ESCO installs separate control equipment in the field and directly manages at least some of the DER. There is close collaboration with the DSO. This function is fully applicable in the liberalized market model of Section 1.5.2.
- **Dedicated microgrid operator:** This is a special configuration suitable for an independent (privately owned) part of the distribution network, such as a mall or an airport. In this case, a dedicated microgrid operator can be responsible for the management of this part of the network. A typical case is the prosumer consortium model of Section 1.5.2.

2.5 Centralized and Decentralized Control

The microgrid structure depicted in Figure 2.4 can be operated in a centralized or decentralized way, depending on the responsibilities assumed by the different control levels. In centralized control, the main responsibility for the maximization of the microgrid value and the optimization of its operation lies with the MGCC. The MGCC using market prices of electricity and gas costs, and based on grid security concerns and ancillary services requests by the DSO, determines the amount of power that the microgrid should import from the upstream distribution system, optimizing local production or consumption capabilities. The defined optimized operating scenario is realized by controlling the microsources and controllable loads within the microgrid by sending control signals to the field. In this framework, non-critical, flexible loads can be shed, when profitable. Furthermore, it is necessary to monitor the actual active and reactive power of the components. In a fully decentralized approach, the main responsibility is given to the MCs that compete or collaborate, to optimize their production, in order to satisfy the demand and probably provide the maximum possible export to the grid, taking into account current market prices. This approach is suitable in cases of different ownership of DERs, where several decisions should be taken locally, making centralized control very difficult. Apart from the main objectives and characteristics of the controlled microgrid, the choice between centralized and decentralized approaches for microgrid control, depends on the available or affordable resources: personnel and equipment. The two approaches are presented in Figures 2.8 and 2.9. In both approaches, some basic functions are centrally available, such as local production and demand forecasting and security monitoring.

In this section, the effectiveness of the two approaches in terms of calculation time, scalability and accuracy is discussed. At this point, the introduction of algorithm complexity is useful. Algorithm complexity refers to the calculation time that an algorithm needs to finish a task. This calculation usually correlates the execution time needed with one or more arguments, for example, the solution time of the unit commitment problem is correlated to the number of the generating units available. Capital O is used to denote the algorithms classification according to their response (e.g. processing time, number of messages exchanged or memory requirements) to changes in input size. As an example, let's consider the problem of sorting a list of n numbers into ascending order. The algorithm sorts the list by swapping at each iteration a pair of numbers between two positions of the list. The critical task in this algorithm is the number of swaps between numbers. If the complexity of the sorting

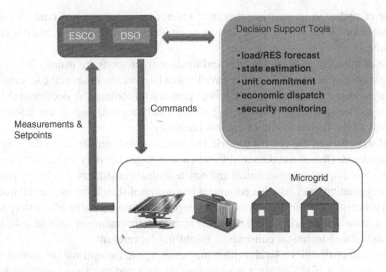

Figure 2.8 Principles of centralized control

algorithm is $O(n^2)$, this means that the maximum number of swaps is the square of the number of items in the list, so a list of 10 numbers requires at maximum 100 swaps, a list of 100 numbers requires at maximum 10 000 swaps, and so on. This notion is important in order to compare the two approaches not only with respect to the number of nodes (e.g. number of items in the list), but also their correlation with the most time- or effort-consuming action (e.g. number of swaps). Next, the key attributes that affect the performance of the control algorithms for microgrids are listed.

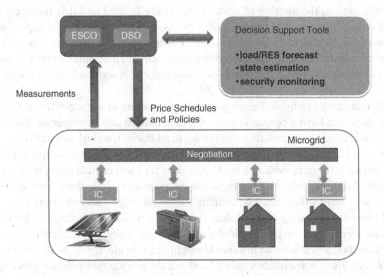

Figure 2.9 Principles of decentralized control

- **Number of nodes:** A microgrid consists of several microsources and controllable loads. The number of DERs affects critically the complexity of the problem and the computational time.
- **Number of message exchanges:** DGs and loads in microgrids are usually dispersed, and communication systems at LV usually have limited bandwidth. In several cases the number of messages required to perform a task is of primary importance. A decentralized control approach reduces the number of messages, since only a small part of the information is transferred to the higher levels of control hierarchy.
- **Size and structure of the system model:** The structure and complexity of the system need to be considered. Decisions taken by different actors might not only increase the number of nodes, but also impose extra technical and non-technical constraints. A relevant issue is the level of information, such as what parameter or constraint should be made available for the decision process of the various actors. For example, the state of charge of a battery might be important to the neighboring DGs, while the internal temperature or voltage level of a cell and the associated technical constraints, might not be relevant.
- **Accuracy and optimality:** An algorithm may converge to the optimal solution or near the optimal solution. It is self-evident that the convergence and the accuracy of the solutions depend on the accuracy of the models used and of the relevant input data. The question is whether a suboptimal solution is acceptable and, if so, at what cost.

The choice between the centralized and decentralized approach for microgrid control, depends on the main objectives and the special characteristics of the controlled microgrid and the available or affordable resources: personnel and equipment.

Clearly, centralized control is more suitable if the users of the microgrid (DG and load owners) have common goals or a common operational environment seeking cooperation, in order to meet their goals. Such an example is an industrial microgrid, in which a single owner might exercise full control of all its energy sources and loads, is able to continuously monitor them and aims to operate the system in the most economical way. Considering the general attributes listed earlier, the number of nodes is generally limited and it is relatively easy to install a fast communication system and a set of sensors. Dedicated operating personnel for the operation of the microgrid might be available. Furthermore, the optimization problem has a limited set of constraints and specific objectives, such as cost minimization. Finally, the requested solution should be as accurate as possible since a suboptimal solution may lead to profit losses.

Microgrids operating in a market environment might require that competitive actions of the controller of each unit have a certain degree of independence and intelligence. Furthermore, local DER owners might have different objectives: in addition to selling power to the network, they might have other tasks, such as producing heat for local installations, keeping the voltage locally at a certain level or providing a backup system for local critical loads in case of main system failure. Some microgrid customers might primarily seek their own energy cost minimization and have diverse needs, although they all might benefit from the common objective of lowering operating costs of their feeder. In a residential microgrid, for example, one household might have at one particular moment increased electric energy needs, for example for cooking, while another household might need no electricity at all, because all its tenants are absent. Both households would like to sell the extra power produced locally to the grid, but it is unlikely that they would accept remote control of their production. Considering again the general attributes mentioned earlier, we should identify that in this case the number of nodes might increase significantly. A neighborhood might have dozens of households or installed DGs, and if we consider multi-microgrids the number is increasing further. In such cases, it is not possible to have a dedicated communication system, but existing infrastructure should be used. Thus, part of the system might not have sufficient bandwidth or the communication could be expensive. An approach that limits the amount of data transfer is essential. The availability of powerful computing facilities or dedicated operators is also highly unlikely.

Furthermore, the optimization problem is becoming extremely complex, as a result of specific characteristics. For example, it is extremely complicated to model the comfort requirements in each household or to include all the special technical constraints of all appliances in a single optimization problem. The decentralized approach suggests that this type of constraint and sub-problem should be solved locally in each household or DG. In the general control problem, each household could be presented as a load node that has the ability to shed or shift some load, or a production node with DGs that can offer a certain amount of energy without caring about the type of the engine or the technical constraints. Finally, in this case, a suboptimal solution is probably acceptable, given the high costs of the installation of fast communication networks or powerful processors dedicated to energy optimization.

Another important factor is the openness of the system. The distributed computing technology allows DG manufacturers and loads to provide plug-and-play components by embedding control agents (following some rules) in their devices. The software should be intelligent enough to monitor the process and follow the best policy. The availability of dedicated personnel responsible for system installation and maintenance, likely in centralized systems, might relax this requirement to some extent. In this case, dedicated personnel could also be available for monitoring the process, which could intervene in an emergency.

The general conclusion is that the centralized approach is suitable for a system with one specific goal and a decentralized one in a system with several goals.

The above considerations are summarized in the Table 2.1.

2.6 Forecasting

2.6.1 Introduction

Both centralized and decentralized control approaches require forecasts of the electricity demand, heat demand, generation from renewable power sources and external electricity prices, for the next few hours, as shown in Figures 2.8 and Figure 2.9. Forecasting the evolution of these quantities allows us to face unsafe situations and to optimize production costs and in general to maximize revenues of the production process in the marketplace. As a consequence, forecasting options may have a direct impact on the economic viability of microgrids, since they allow them to enhance their competitiveness compared to centralized generation.

The aim of this section is to introduce the problem of short-term forecasting in the frame of microgrids and to provide example methods for forecasting relevant quantities. *It is recognized, however, that it is still premature to propose a particular operational forecasting tool for microgrids.* First, the role of forecasting functionalities in microgrids needs to be discussed.

Table 2.1 Considerations in the applicability of centralized and decentralized control. Reproduced by permission of the IEEE

	Centralized control	Decentralized control
DG ownership	Single owner	Multiple owners
Goals	A clear, single task, e.g. minimization of energy costs	Uncertainty over what each owner wants at any particular moment
Availability of operating personnel (monitoring, low-level management, special switching operations, etc.)	Available	Not available
Market participation	Implementation of complicated algorithms	Owners unlikely to use complex algorithms
Installation of new equipment	Requirements of specialized personnel	Should be plug-and-play
Optimality	Optimal solutions	Mostly suboptimal solutions
Communication requirements	High	Modest
Market participation	All units collaborate	Some units may be competitive
Microgrid operation is attached to a larger and more critical operation	Possible	Not possible

2.6.1.1 Are Forecasting Functionalities Relevant for Microgrids?

Depending on the mode of operation, it is clear that in islanded operation, prediction of demand is of primary importance, since the aim is to achieve the balance of the system. In interconnected operation, however, the importance of demand or production forecasting may change, depending on the focus – a system-driven or a customer-driven approach. In the first case, forecasting functions may have less importance since we may consider that a microgrid connected to an "infinite" source of power is able to cover any deficit at any time. However, in a customer-driven approach, economics – and thus forecasting – gain in importance. If microgrids is the "business case" of an energy service provider, who has to consider electricity prices, then decisions will be based on forecasting. Forecasting functions gain in importance when one considers multi-microgrids scenarios.

The scale of a microgrid imposes the consideration of cost-effective approaches for forecasting. Today, forecasting technology for renewable generation is not plug-and-play. Developing and implementing forecasting options for a power system application involves costs for research and development, instrumentation for data collection, operational costs for numerical weather predictions and so on. Forecasts can be provided commercially, either in the form of a service or by software installed on-site. In any case, forecasting solutions have a price that should be compared to the benefits they provide. Decentralizing power generation, especially by adding renewables, adds intermittency in power generation, and retaining the same quality of service means that accurate forecasting is a cost-effective solution to counterbalance intermittency. Another relevant issue is the acceptability of power system operators, who are used to manage almost deterministic processes; for example, they are able, even without mathematical tools, to forecast the system load with an impressive accuracy of a few percent. The capacity to accept more intermittent options is linked to the capacity to provide tools to compensate intermittency and manage uncertainties. This is especially true, in electricity market operation, where penalties are associated to uncertainties, and decisions have to be taken as a function of the prices in the near future.

Considerable work in the power systems area has been devoted to forecasting demand, wind power, heat demand and, more recently, electricity prices and PV generation. The work, especially on demand, concerns mainly large interconnected systems. Less experience is available on forecasting for smaller systems and with a high temporal resolution (i.e. 5–10 minutes) for the next 1–4 hours. For this reason, in very small applications persistence is usually applied. This is a simple method saying that the predicted variable will retain the current value during the next period:

$$\hat{P}(t + k) = P(t), \quad k = 1, 2, \ldots n \tag{2.1}$$

This can be the baseline model for heat, wind and price forecasting, while for load or PV generation one could use the "diurnal persistence" which consists in using as forecast the measured value of the process at the same hour of the previous day. Using this model for taking decisions may reduce benefits, especially in electricity markets with highly volatile prices. In order to quantify the value of forecasting, we need to simulate the operation of a microgrid using persistence against perfect forecasting. The difference between the two would indicate the interest in investing in advanced forecasting methods. Results from such a study would, however, be difficult to generalize, since they depend critically on the structure of the microgrid and the characteristics of the electricity market.

2.6.2 Demand Forecasting

In interconnected or large island power systems, demand depends on weather conditions, habits and activities of the customers and thus it is highly correlated to the time of day and the type of the day or season of the year. Predictions are usually required for the next 24/48 hours in hourly or 30-minute time steps. Typically, forecasting accuracy is high – in the order of 1–5% – depending on the time horizon and the type/size of the system. Uncertainty can be estimated by classical methods, such as resampling. A large number of methods for load forecasting can be found in the literature, for example, an extended review focusing on artificial intelligence based techniques is given in [5,6].

Downscaling the demand prediction problem to smaller power systems, such as the systems of islands, increases the difficulties, because the variability of the load also increases. At the level of a microgrid, the aggregation or smoothing effect is significantly reduced, and uncertainty increases as the size of the system gets smaller. To meet this difficulty we should add the increase in time resolution. We enter the area of very-short-term forecasting with reduced smoothing effects [7]. However, at the level of a feeder, the aggregation is still enough so that time-series approaches can be applied for forecasting. When it gets down to the level of a single client (i.e. demand of a house) there is, in general, a lack of measured data to adequately characterize the problem. The deployment of smart meters permits the collection of data for that purpose.

The shortest time resolution for load forecasting in microgrids could not be less than 5–10 minutes, if one would like to speak about large-scale applications. The load patterns of individual customers can be highly correlated with each other, that is, due to external temperature dependency. First attempts to predict customer load using smart meter data show that such data have a quite high variability, but there is still a prominent daily pattern that makes the time series predictable. First evaluation results suggest an accuracy at the level of 30% of mean absolute percentage error. In contrast to the classical load forecasting problem, it is also expected that the demand will be correlated to electricity prices, especially when customer behavior is influenced by dynamic demand-side management actions. Prediction models for demand may consider electricity prices as input (predictions) to accommodate this correlation.

2.6.2.1 Contribution of Weather Predictions to Operation at Best Efficiency Point

The consideration of weather forecasts as a general input to the various forecasting functions in the MGCC seems to be an option that may be exploited in multiple ways. Apart from their use for reliable forecasting of production of renewable units and power demand, the role of weather forecasts can be also important in predicting the operation (efficiency) of micro turbines. This aspect is also important in larger-scale systems or multi-microgrid systems. For light load conditions, it would be better to have fewer microturbines running, but running at rated load, than to have several microturbines running at partial load. This is because microturbines are more efficient when operating at rated load. The decision of how many machines to run, and at what load, can best be made by the MGCC, because it has knowledge of the process condition, the weather forecast and the production schedule. This requirement is also identified in [2] for energy management systems destined for microgrids.

2.6.3 Wind and PV Production Forecasting

Although for microgrids developed in urban environments, wind energy might not be widely adopted, small wind turbines can provide a viable option with a high potential in several cases. Short-term forecasting is of primary importance for integrating wind energy, especially in larger power systems, and there is very rich literature on the subject. Research in wind power forecasting is a multidisciplinary field, since it combines areas such as meteorology, statistics, physical modeling, computational intelligence [8]. Similar efforts exist in other areas, such as forecasting of PV, heat and hydro. An excellent reference with a detailed state of the art on wind power forecasting is provided in [9]. In the case of very-short-term wind power prediction a review of available models is given in [10].

In microgrids, forecasts of renewable generation can be provided in a centralized way (i.e. in the case of an MGCC) using input from weather forecasts and past measurements. In a decentralized management approach, where local intelligence has to be considered (i.e. at the level of customers with PV panels and batteries), plug-and-play approaches that use basic weather forecasts from the internet, or simply measurements, can be considered. The plug-and-play capacity refers to the requirement for low human intervention and also to the possibility of providing forecasts of adequate accuracy in cases with little history of measured data available, as can be the case with new PV plant installations. Both physical methods and statistical methods, such as fuzzy neural networks, random forests, regime switching and kernel density estimators, can be used.

2.6.4 Heat Demand Forecasting

Forecasting heat consumption is a necessary functionality for the MGCC of a micro-CHP based microgrid. The main factors affecting heat demand are:

- time of day effect,
- weekend/weekday effect,
- seasonal effects,
- time varying volatility,
- high negative correlation between heat demand and external temperature.

Several approaches have been developed for online prediction of heat consumption in district heating systems. The time horizon is often 72 hours and the time step hourly.

The simpler approaches are based on purely autoregressive moving average (ARMA) models that use only heat demand data as input. Models considering seasonal differencing have been also applied. As an extension, models are also considered with temperature as an explanatory variable. More advanced developments assume that meteorological forecasts are available online, although such a facility is not commonly expected in microgrids.

The methods of prediction applied are based on adaptive estimation that allows for adaptation to slow changes in the system. This approach is also used to track the transition from, say, warm to cold periods. Due to different preferences of the households to which the heat is supplied, this transition is smooth.

Alternatively, black box models such as neural networks [11,12] or fuzzy logic neural networks can be applied. In this case, more flexibility is gained regarding the structure of the model and the available information according to the application.

2.6.5 Electricity Prices Forecasting

Short-term forecasting of electricity prices may be important in volatile electricity markets. Spot prices may significantly influence decisions on the use of microsources. Various approaches have been tested for this purpose. Electricity prices vary from other commodities because the primary good, electricity, cannot be stored, implying that inventories cannot be created and managed to arbitrage prices over time. As an example, the process in the Leipzig Power Exchange can be characterized by the following features [12,13]:

- strong mean reversion: deviation of the price due to random effects are corrected to a certain degree,
- time of the day effect,
- calendar effects such as weekdays, weekends and holidays,
- seasonal effects,
- time-varying volatility and volatility clustering,
- high percentage of unusual prices, mainly in periods of high demand,
- inverse leverage effect: a positive price shock has a bigger impact than a negative one,
- non-constant mean and variance

Models applied for short-term price forecasting include [14]:

- mean reverting processes
- mean reverting processes with time-varying mean
- autoregressive moving average models (ARMA)
- exponential generalized autoregressive conditional heteroscedasticity models (EGARCH).

2.6.6 Evaluation of Uncertainties on Predictions

In general, statistical methods based on machine learning are among the promising approaches that can be applied for forecasting purposes in microgrids. However, forecasting models, especially of load or heat demand, have to be validated with measured data reflecting the situation in real microgrids. As discussed above, due to the small size of microgrids, the smoothing effect is reduced due to the limited aggregation of the forecasted processes. Moreover, the need for higher time resolution results in an increase of intermittency. Therefore, in parallel to the research on forecasting models, it is necessary to develop research for online evaluation of the uncertainty of the predictions. Such approaches for wind power forecasting in larger systems have been developed, and it is today possible to provide directly probabilistic forecasts for wind or PV prediction. These provide the whole distribution for each time step and from this one can obtain various uncertainty products such as quantiles or prediction intervals [15]. Studying the predictability and the variability of the various processes related to microgrids is of major importance for deciding what kind of approaches are appropriate for the management functions (i.e. deterministic or probabilistic ones) [16]. The development of cost-effective prediction tools with plug-and-play capabilities suitable for the limited facilities of a microgrid environment is still an open research issue.

2.7 Centralized Control

Microgrids can be centrally managed by extending and properly adapting the functionalities of existing energy management system (EMS) functions. Regarding steady-state operation, as shown in Figure 2.8, the basic feature of centralized control is that decisions about the operation of the DER are taken by the microgrid operator or ESCO at the MGCC level. The MGCC is equipped, among other things, with scheduling routines that provide optimal setpoints to the MCs, based on the overall optimization objectives. This section describes the scheduling functions that are required for centralized scheduling of microgrids [17].

The local distributed energy sources, acting either as individual market players or as one coordinated market player, provide energy and ancillary services by bidding in energy and ancillary markets, based on the prices provided by the system. Two market policies can be distinguished: In the first case, the microgrid serves its own needs only, displacing as much energy from the grid as economically optimal. In the second case, the microgrid participates in the market probably through an energy service provider or aggregator. Due to its size and the uncontrollability of the microsources, it is unlikely that the microgrid bids will concern longer term horizons. It is conceivable, however, to have microgrid bids covering a short time ahead, say the next 15–30 minutes.

Moreover, in normal interconnected operation, individual consumers can participate in the market operation, providing load flexibility, directly or indirectly, by suitable programmable controllers. It is assumed that each consumer may have low and high priority loads and would

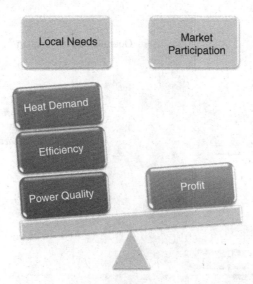

Figure 2.10   The control system should be able to balance between market participation and local needs

send separate bids to the MGCC for each of them. In this way, the total consumption of the consumer is known in advance. Some of the loads will be served and others not, according to the bids of both the consumers and the local power producers. Two options can be considered for the consumers' bids: (a) the consumer's bid for supply of high and low priority loads or (b) the consumer's offer to shed low priority loads at fixed prices in the next operating periods. For the loads that the MGCC decides not to serve, a signal is sent to the load controllers in order to interrupt the power supply.

It should be noted that the owners of DGs or flexible loads might not have, as a primary motivation, profit maximization obtained in the wholesale market. Instead, their goal might be to satisfy other needs, such as heat demand or increased quality of service (power quality). The control system should be able to identify the specific needs in each case and to use the market services in the most beneficial way (Figure 2.10). The balance between individual needs and market participation should be found in each case, separately.

2.7.1 Economic Operation

A typical microgrid operates as follows: The local controller MC takes into account the operational cost function of the microsource, a profit margin sought for by the DG owner, and the prices of the external market provided by the MGCC, in order to announce offers to the MGCC as well as technical constraints. These offers are made at fixed time intervals m for the next few hours, that is, the optimization horizon. A typical interval might be 15 minutes, if we assume system operation in line with the functions of current AMR/AMI systems. The MGCC optimizes the microgrid operation according to the external market prices, the bids received by the DG sources and the forecasted loads, and sends signals to the MCs of the DG sources to be committed and, if applicable, to determine the level of their production. In addition, consumers within the microgrid might bid for supply of their loads for the next hour in the

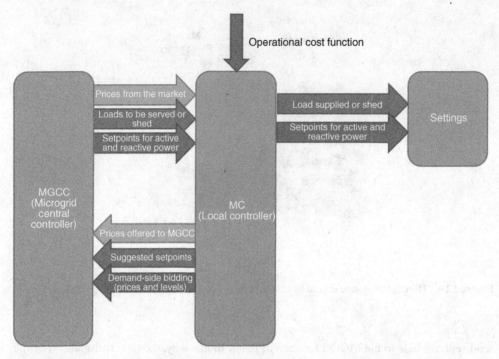

Figure 2.11 Closed loop for energy markets – information exchange diagram

same m minute intervals, or might bid to curtail their loads, if fairly remunerated. In this case, the MGCC optimizes operation based on DG sources and flexible load bids, and sends dispatch signals to both the MCs and LCs. Figure 2.11 shows a typical information exchange flow in microgrid operation.

The optimization procedure clearly depends on the market policy adopted in the microgrid operation. In the following section, alternative market policies are considered.

2.7.2 Participation in Energy Markets

2.7.2.1 Market Policies

Two market policies are assumed: In the first policy the MGCC aims to serve the total demand of the microgrid, using its local production, as much as possible when financially beneficial, without exporting power to the upstream distribution grid. Moreover, the MGCC tries to minimize its reactive power requests from the distribution grid. This is equivalent to the "good citizen" behavior, as termed in [18]. For the overall distribution grid operation, such a behavior is beneficial, because:

- at the time of peak demand leading to high energy prices in the open market, the microgrid relieves possible network congestion by supplying partly or fully its energy needs
- the distribution grid does not have to deal with the reactive power support of the microgrid, making voltage control easier.

From the end-users point of view, the MGCC minimizes operational cost of the microgrid, taking into account open market prices, demand and DG bids. The end-users of the microgrid share the benefits of reduced operational costs.

In the second of the two policies, the microgrid participates in the open market, buying and selling active and reactive power to the grid, probably via an aggregator or energy service provider. According to this policy, the MGCC tries to maximize the value of the microgrid, that is, maximize the corresponding revenues of the aggregator, by exchanging power with the grid. The end-users are charged for their active and reactive power consumption at open market prices. From the grid's point of view, this is equivalent to the "ideal citizen" behavior referred to in [18]. The microgrid behaves as a single generator capable of relieving possible network congestion not only in the microgrid itself, but also by transferring energy to nearby feeders of the distribution network.

It should be noted that the MGCC can take into account environmental parameters such as green house gas (GHG) emissions reductions, optimizing the microgrid operation accordingly.

2.7.2.2 Demand Side Bidding

Each consumer may have low and high priority loads allowing him to send separate bids to the MGCC for each type of load. Without loss of generality, it is assumed that each consumer places bids for his demand at two levels, and the prices reflect his priorities. It is preferable that "low" priority loads are not served when the market prices are high, but can either be satisfied at periods of lower prices (shift) or not served at all (curtailment). Two options are considered for the consumers' bids:

A. Shift option
   Consumers place two different bids for the supply of their high and low priority loads.
B. Curtailment option
   Consumers offer to shed low priority loads at fixed prices in the next operating periods being remunerated for this service.

In both options, the MGCC

- informs consumers about the external market prices,
- accepts bids from the consumers every hour, corresponding to m minute intervals, and
- sends signals to the MCs, according to the outcome of the optimization routine.

The external market prices help consumers prepare their bids. According to the "good citizen" policy, these prices correspond to the highest prices that the end-users can possibly be charged, if security constraints are not considered. The MGCC optimizes the microgrid operation according to the bids of both DG and loads. In the shift option, the MGCC sums up the DG sources' bids in ascending order, and the demand side bids in descending order, in order to decide which DG sources will operate for the next hour and which loads will be served. This is shown schematically in Figure 2.12. Optimal operation is achieved at the intersection point of the producers and demand bids.

In the curtailment option, consumers bid for the part of their load that they are willing to shed in the next time intervals, if compensated. A possible formulation of the customer bid is

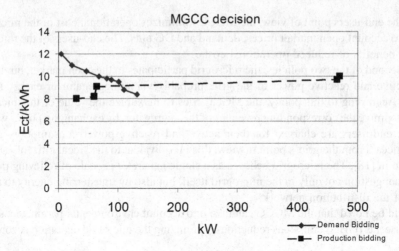

Figure 2.12 The decision made by the MGCC for the shift option. Reproduced by permission of the IEEE

shown in Figure 2.13. The main difference with the shift option is that the MGCC knows the current total demand of the microgrid, and sends interruption signals to the MCs, if financially beneficial.

2.7.2.3 Security Issues

Similar to large power systems, steady-state security issues concern operation of the microgrid satisfying voltage constraints and power flows within thermal limits. A critical consideration concerns overloading the interconnection between the microgrid with the upstream distribution network. Dynamic security issues could also be considered to ensure microgrid operation under a number of contingencies within and above it. For microgrids the seamless transition between interconnected and islanded mode of operation is of particular importance. Such

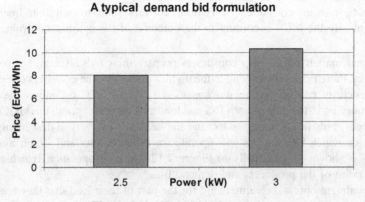

Figure 2.13 Typical bid formulation

security considerations can be expressed as additional constraints and might affect the optimization outcome [19].

2.7.3 Mathematical Formulation

2.7.3.1 General

The optimization problem is formulated differently, according to the market policies assumed. Since there are no mature reactive power markets at the distribution level, such a market is not considered within a microgrid. If, however, such a market is to be implemented, the following functions can be easily altered to take it into account.

2.7.3.2 Market Policy 1

As discussed in Section 2.7.2, the MGCC aims to minimize the microgrid operational cost. It is assumed that the operator of the MGCC is a non-profit organization and the end-users share the benefits. The scope is to lower electricity prices for the microgrid end-users and protect them, as much as possible, from the volatility of the open market prices. The objective function to be minimized for each one of the m-minute intervals is

$$\text{cost} = \sum_{i=1}^{N} \text{active\_bid}(x_i) + AX + \sum_{j=1}^{L} \text{load\_bid}(bid_j) \qquad (2.2)$$

active_bid(x_i) is the bid for active power from the i-th DG source.
x_i is the active power production of the i-th DG source.
X is the active power bought from the grid.
N is the number of the DG sources that offer bids for active power production.
A is the price on the open market for active power.

If demand side bidding is considered, then bid_j refers to the bid of the j-th load of the L loads bidding. If the customer is compensated, then the cost of compensation load_bid(bid_j), assumed as a linear function of bid_j, should be added to the operation cost.

The constraints for this optimization problem are

- technical limits of the DG sources, such as minimum and maximum limits of operation, P–Q curves and start-up times
- active power balance of the microgrid, Eq. (2.3), where P_demand is the active power demand.

$$X + \sum_{i=1}^{N} x_i + \sum_{j=1}^{L} bid_j = \text{P\_demand} \qquad (2.3)$$

2.7.3.3 Market Policy 2

According to this policy, the MGCC (aggregator) maximizes revenues from the power exchange with the grid. End-users are assumed to be charged with open market prices.

The optimization problem is

$$\text{Maximize}\{\text{Income} - \text{Expenses}\} = \text{Maximize}\{\text{Revenues}\}$$

Income comes from selling active power to both the grid and the microgrid end-users. If the demand is higher than the production of the DG sources, power bought from the grid is sold to the end-users of the microgrid. If the demand is lower than the production of the DG sources, term X in Eqs. (2.4) and (2.5) is zero.

$$\text{Income} = AX + A \cdot \sum_{i=1}^{N} x_i \tag{2.4}$$

The term Expenses includes costs for active power bought from the grid plus compensation to DG sources. If demand side bidding is considered, relevant costs are added to Expenses.

$$\text{Expenses} = \sum_{i=1}^{N} \text{active\_bid}(x_i) + AX + \sum_{j=1}^{L} \text{load\_bid}(\text{bid}_j) \tag{2.5}$$

The MGCC should maximize Eq. (2.6)

$$\text{Revenues} = A \cdot \sum_{i=1}^{N} x_i - \sum_{i=1}^{N} \text{active\_bid}(x_i) \\ - \sum_{j=1}^{L} \text{load\_bid}(\text{bid}_j) \tag{2.6}$$

Constraints are the technical limits of the units and that at least the demand of the microgrid should be met, as expressed by Eq. (2.7).

$$X + \sum_{i=1}^{N} x_i + \sum_{j=1}^{L} \text{bid}_j \geq \text{P\_demand} \tag{2.7}$$

2.7.4 Solution Methodology

There are several methods for solving the unit commitment (UC) problem, namely to identify which of the bids of both DGs and loads will be accepted. A simple method is the use of a priority list. The DG bids, the load bids, if DSB options are applied, and the external market prices are placed sequentially according to their differential cost at the highest level of production for the specific period. This list is sorted in ascending bid values so that the total demand is met.

DG bids are assumed to have a quadratic form:

$$\sum_{i=1}^{N} \text{active\_bid}(x_i) = a_i x_i^2 + b_i x_i + c_i \tag{2.8}$$

a_i is the quadratic coefficient of the bids for active power bid.
b_i is the linear coefficient of the bids for active power bid.
c_i is the constant term coefficient of the bids for active power bid.

Economic dispatch (ED) must be performed next, so that the production settings of the DG sources, whose output can be regulated, and the power exchange with the grid are determined. The production of non-regulated DG and the loads that will not be served has been decided by the UC function, as described in the previous paragraph. If the bids are continuous convex functions, like Eq. (2.8) then mathematical optimization methods can be utilized, such as sequential quadratic programming (SQP), as described in [20]. Artificial intelligence techniques can also be used, especially if scalar or discontinuous bids are considered [21,22]. The rest of the demand is met by the DG sources and the power bought from the grid.

2.7.5 Study Case

Results from a typical study case LV network, shown in Figure 2.14 [23], are presented in this section. Network data and the other parameters of the study are included for completeness. The network comprises three feeders, one serving a primarily residential area, one industrial feeder serving a small workshop and one commercial feeder. Load curves for each feeder and the whole microgrid for a typical day are shown in Figure 2.15. The total energy demand for

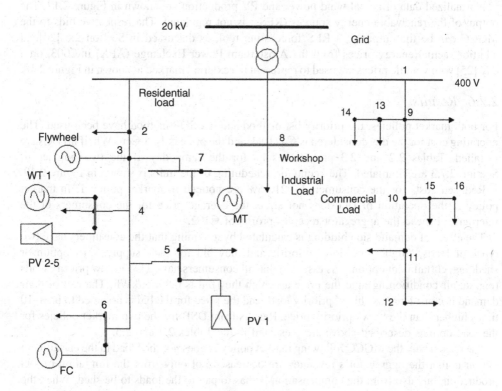

Figure 2.14 The study case LV network. Reproduced by permission of the IEEE

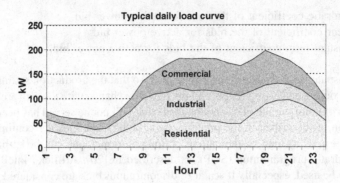

Figure 2.15 Typical load curve for each feeder of the study case network. Reproduced by permission of the IEEE

this day is 3188 kWh. The power factor of all loads is assumed to be 0.85 lagging. A variety of DG sources, such as a microturbine (MT), a fuel cell (FC), a directly coupled wind turbine (WT) and several PVs, are installed in the residential feeder, as shown in Figure 2.16. It is assumed that all DG produce active power at unity power factor. The resistances and reactances of the lines, the capacity of the DG sources and their bids parameters, installation costs and basic economic assumptions are provided in Appendix 2.A.

Normalized data of actual wind power and PV production are shown in Figure 2.17. The output of the renewable energy sources (RESs) is not regulated. The respective bids to the MGCC can be the output of a RES forecasting tool, as discussed in Section 2.6 [24]. In addition, actual energy prices from the Amsterdam Power Exchange (APX) in 2003, on a day [25] with volatile prices are used to represent the external market, as shown in Figure 2.18.

2.7.6 Results

For both market policies, the priority list method and the SQP method have been used. The operating cost for the day considered is €471.83, and the price is 14.8 €ct/kWh, if no DGs are installed. Tables 2.2 and 2.3 provide results for the same day, if the two policies of Section 2.7.3 are simulated. The economic scheduling of the units is shown in Figure 2.19.

Reduced costs for the consumers by 21.56% are noticed in market policy 1. In market policy 2, the operation of DG does not affect the average price for the consumers of the microgrid; instead the aggregator receives profits of €102.

The effect of demand side bidding is calculated by assuming that the consumers have two types of loads, "high" and "low" priority, and they bid for their supply, shift option or shedding, curtailment option. It is assumed that all consumers have 2 kW of low priority loads (e.g. an air conditioning) and the price at which they bid is 6.8 €ct/kWh. The rest of their demand is considered as "high" priority load, and the price for the bid is assumed to be 8–10 times higher than the "low" priority price. Results from DSB for the two market policies for the load options described above are presented in the Tables 2.4 and 2.5.

The reason why the MGCC following market policy 2, does not shed load in the curtailment option is that the aggregator's revenues are decreased not only from the limitation of DG production, but also from the compensation he has to pay to the loads to be shed. When the load shift option is utilized, the revenues of the aggregator do not change, since the energy

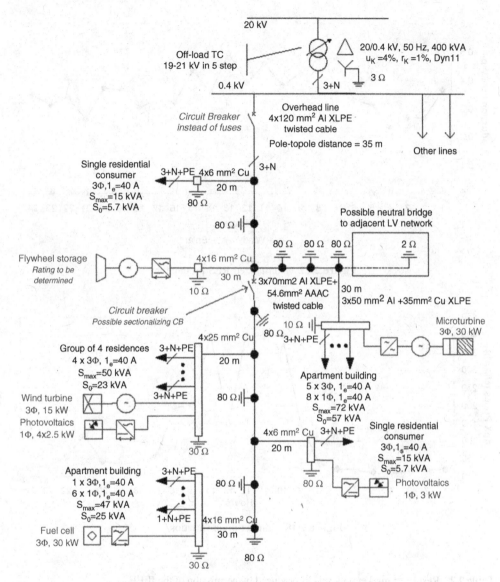

Figure 2.16 The residential feeder with DG sources

produced by the DG sources is sold to the external market at the same prices as in the microgrid internal market. However, the power exchange with the grid is altered by decreasing the grid demand. This service, especially during hours of stress, can be extremely beneficial even for customers that are not part of the microgrid.

This example shows the potential benefits provided by the coordinated operation of DER in a microgrid. Exploiting local DER can significantly reduce costs for the microgrid consumers, or provide revenues to the microgrid's operator, especially in periods of high external market prices. A more complete analysis of microgrid benefits is provided in Chapter 7.

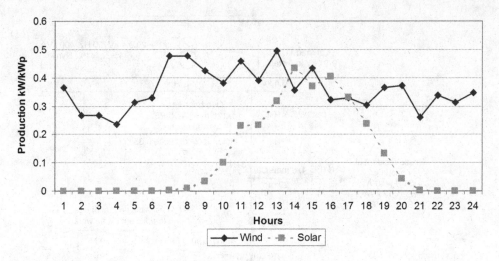

Figure 2.17 Normalized RES production

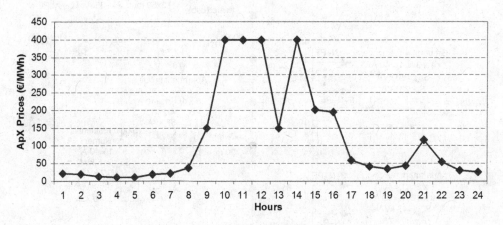

Figure 2.18 Market price variation

Table 2.2 Results of market policy 1. Reproduced by permission of the IEEE

Cost euro	Difference with actual operation	Average price(€ct/kWh)
370.09	21.56%	11.61

Table 2.3 Results of market policy 2. Reproduced by permission of the IEEE

Revenues euro	Percentage of revenues	Average price(€ct/kWh)
101.73	21.56%	14.8

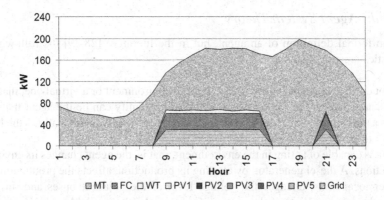

Figure 2.19 Typical results of the daily operation. Reproduced by permission of the IEEE

Table 2.4 Results of market policy 1 with demand side bidding. Reproduced by permission of the IEEE

	Shift option – market policy 1	Curtail option – market policy 1
Revenues (€)	307.66	323.44
Load shed (kWh)	232	232
Cost reduction (%)	34.79	31.44
Average price (€ct/kWh)	10.41	10.94

Table 2.5 Results of market policy 2 with demand side bidding. Reproduced by permission of the IEEE

	Shift option – market policy 2	Curtail option – market policy 2
Revenues (€)	101.73	101.73
Load shed (kWh)	232	0
Revenues (%)	21.56	21.56
Average price (€ct/kWh)	14.8	14.8

2.8 Decentralized Control

The idea of decentralized control is becoming popular nowadays, not only for microgrids but also for other functions of power systems [26,27]. An interesting approach to designing and developing decentralized systems is based on multi-agent system (MAS) theory. The core idea is that an autonomous control process is assumed by each controllable element, namely inverters, DGs or loads. The MAS theory describes the coordination algorithms, the communication between the agents and the organization of the whole system. Practical applications of these technologies are presented in Sections 6.2.1 and 6.2.2. Next, after a short introduction to MAS theory, these three topics will be addressed.

2.8.1 Multi-Agent System Theory

There is no formal definition of an agent, but in the literature [28,29] the following basic characteristics are provided:

- An agent can be a physical entity that acts in the environment or a virtual one, that is, with no physical existence. In our application, a physical entity can be the agent that directly controls a microturbine or a virtual one, such as a piece of software that allows the ESCO or the DSO to participate in the market.
- An agent is capable of acting in the environment, that is, the agent changes its environment by its actions. A diesel generator, by altering its production, affects the production level of the other local units, changes the voltage level of the adjacent buses and, in general, changes the security level of the system, for example, the available spinning reserve.
- Agents communicate with each other, and this could be regarded as part of their capability for acting in the environment. As an example, let's consider a system that includes a wind generator and a battery system: the battery system uses energy from the wind turbine to charge it and it discharges in times of no wind. In order to achieve this operation optimally, the two agents have to exchange messages. This is considered a type of action, because the environment is affected by this communication differently from the case the two agents were acting without any kind of coordination.
- Agents have a certain level of autonomy, which means that they can take decisions without a central controller or commander. To achieve this, they are driven by a set of tendencies. For a battery system, a tendency could be: "charge the batteries when the price for the kWh is low and the state of charge is low, too." Thus, the MAS decides when to start charging, based on its own rules and goals and not by an external command. In addition, the autonomy of every agent is related to the resources that it possesses and uses. These resources could be the available fuel for a diesel generator.
- Another significant characteristic of the agents is that they have partial representation – or no representation at all – of the environment. For example, in a power system the agent of a generator knows only the voltage level of its own bus and it can, perhaps, estimate what is happening in certain buses. However, the agent does not know the status of the whole system. This is the core of the MAS technology, since the goal is to control a very complicated system with minimum data exchange and minimum computational resources.
- Finally, an agent has a certain behavior and tends to satisfy certain objectives using its resources, skills and services. An example of these skills is the ability to produce or store power and an example of the services is the ability to sell power in a market. The way that the agent uses its resources, skills and services characterizes its behavior. As a consequence, it is obvious that the behavior of every agent is formed by its goals. An agent that controls a battery system, and whose goal is to supply uninterruptible power to a load, will have different behavior from a similar battery whose primary goal is to maximize profits by bidding in the energy market.

The agents' characteristics are summarized in Figure 2.20.

2.8.1.1 Reactive versus Intelligent Agents

Any entity or device that has one or more of the characteristics of Figure 2.20 can be considered an agent. But how smart can an agent be?

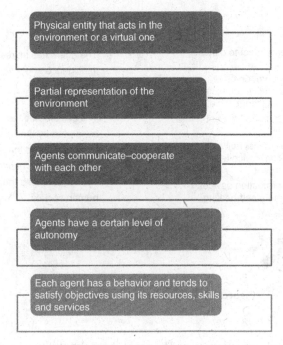

Physical entity that acts in the environment or a virtual one

Partial representation of the environment

Agents communicate–cooperate with each other

Agents have a certain level of autonomy

Each agent has a behavior and tends to satisfy objectives using its resources, skills and services

Figure 2.20 Agents characteristics

Let's consider an under-voltage relay. The relay has the following characteristics:

- has partial representation of the environment (measures the voltage locally only),
- possesses skills (controls a switch), and
- reacts autonomously, according to its goals (opens the switch when the voltage goes beyond certain limits).

According to the literature [28], this type of entity can be considered as a reactive agent that just responds to stimuli from the environment.

What is a cognitive or intelligent agent? Again, there is no formal definition of what an intelligent agent is, but some basic characteristics can be listed:

- memory to acquire and store knowledge of the environment: The memory is one fundamental element of the intelligence and the ability to learn.
- ability to perceive the environment: Having an internal modeling and representation of the environment, detailed enough, is mandatory in order to support the decision making process. The environment model allows the agent to understand the environment state having local information and to predict the effect of a possible action on the environment.
- ability to take decisions according to its memory and the status of the environment and not just to react: The agent has a process (algorithm) that uses the model of the environment and the measurements in order to define the next actions.

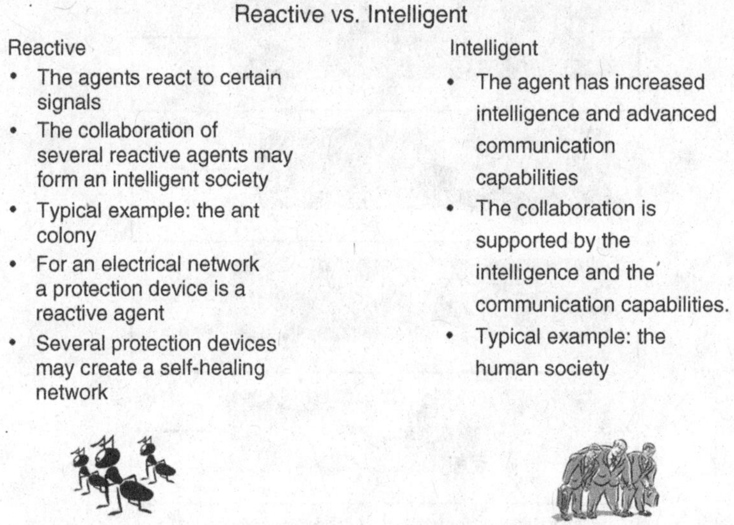

Reactive vs. Intelligent

Reactive
- The agents react to certain signals
- The collaboration of several reactive agents may form an intelligent society
- Typical example: the ant colony
- For an electrical network a protection device is a reactive agent
- Several protection devices may create a self-healing network

Intelligent
- The agent has increased intelligence and advanced communication capabilities
- The collaboration is supported by the intelligence and the communication capabilities.
- Typical example: the human society

Figure 2.21   Reactive and intelligent agents

- ability for high level communication: The agents have the ability to exchange knowledge and use communication as a tool to proceed with complex coordinated actions.

Figure 2.21 summarizes the differences between reactive and intelligent agents, using two well-known societies, human society and an ant colony. In the case of ant colonies, ants do not have significant intelligence and they simply react to stimuli; however, they manage to achieve their main goal, to preserve the society and feed the queen. Human society is formed of intelligent agents: humans. The human intelligence is strengthened by the capability to exchange ideas and knowledge, that is, to communicate.

Finally, we should define the concept of the multi-agent system. An MAS is a system comprising two or more reactive or intelligent agents. It is important to recognize that usually there is no global coordination, simply the local goals of each individual agent are sufficient for a system to solve a problem. Furthermore, under Wooldridge's definitions [30], intelligent agents must have social ability and therefore must be capable of communication with each other.

2.8.2 Agent Communication and Development

Communication is one of the most critical elements that allow the intelligent agents to form a society, a multi-agent system. The transmission of the messages can be done via any traditional communication system, such as IP communications, wired or wireless channels. This section focuses on the context and the structure of the messages. Figure 2.22 presents the agent version of the story of the Tower of Babel, where the workers could not finish the tower due to the lack of communication [31].

This figure presents the main characteristics of the agent communication language (ACL) and the associate problems:

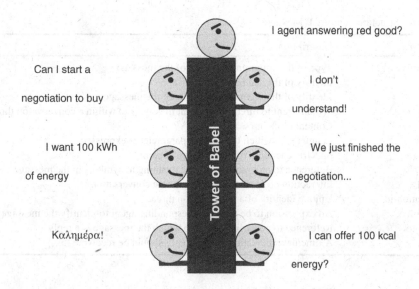

Figure 2.22   The problem of communication. Reproduced by permission of the IEEE

1. The first problem is the ontology or the vocabulary. All the agents speak English except one who says "good morning" in Greek (Καλημέρα!). It is important that the agents should have a common vocabulary. Furthermore, same words should have the same meaning for all agents. In the example, one agent asks for energy in kWh and the other answers in kcal. It is obvious that both agents do not use the word "energy" in the same way.
2. One of the agents says "I agent answering red good?", which is a phrase without an understandable meaning. The agent messages should have a common structure or syntax. In the case of the ACL, each message is actually a set of strings or objects, each one of which has a specific role and meaning.
3. Finally, an agent asks to start the negotiation while the second replies that it has just finished. This is a critical problem in an environment with multiple and parallel dialogs. It is important to understand in which conversation each message belongs and to which question or request it replies.

All these issues are further analyzed in the following sections. Before that, the Foundation of Intelligent Physical Agent and a platform for agent development are introduced.

2.8.2.1   Java Agent Development Framework (JADE)

JADE (Java Agent DEvelopment framework [32]) is a software development framework aimed at developing multi-agent systems and applications conforming to the Foundation of Intelligent Physical Agent's [33] (FIPA's) standards for intelligent agents. FIPA is an IEEE Computer Society standards organization that promotes agent-based technology and the interoperability of its standards with other technologies. FIPA developed a collection of standards that are intended to promote the interoperation of heterogeneous agents and the

Table 2.6   Structure of an ACL Message

Parameter	Description
performative	Type of the communicative act of the message
sender	Identity of the sender of the message
receiver	Identity of the intended recipients of the message
reply-to	Which agent to direct subsequent messages to within a conversation thread
content	Content of the message
language	Language in which the content parameter is expressed
encoding	Specific encoding of the message content
ontology	Reference to an ontology to give meaning to symbols in the message content
protocol	Interaction protocol used to structure a conversation
conversation-id	Unique identity of a conversation thread
reply-with	An expression to be used by a responding agent to identify the message
in-reply-to	Reference to an earlier action to which the message is a reply
reply-by	A time/date indicating when a reply should be received

services that they can represent. The complete set of specifications covers different categories: agent communication, agent management, abstract architecture and applications. Of these categories, agent communication is the core category at the heart of the FIPA multi-agent system model.

JADE includes two main products: a FIPA-compliant agent platform and a package to develop Java agents. JADE has been fully coded in Java, and so an agent programmer, in order to exploit the framework, should code agents in Java, following the implementation guidelines described in the programmer's guide. This guide supposes that the reader is familiar with the FIPA standards, at least with the *Agent Management* specifications (FIPA no. 23), the *Agent Communication Language* and the *ACL Message Structure* (FIPA no. 61) (Table 2.6).

JADE is written in the Java language and is made of various Java packages, giving application programmers both ready-made pieces of functionality and abstract interfaces for custom application-dependent tasks. Java was the programming language of choice because of its many attractive features, particularly geared towards object-oriented programming in distributed heterogeneous environments; some of these features are object serialization, reflection API and remote method invocation (RMI).

The standard model of an agent platform, as defined by FIPA, is presented in Figure 2.23.

The agent management system (AMS) is the agent that exerts supervisory control over access to and use of the agent platform. Only one AMS will exist in a single platform. The AMS provides white-page and life-cycle service, maintaining a directory of agent identifiers (AIDs) and the agent state. Each agent must register with an AMS in order to get a valid AID.

The directory facilitator (DF) is the agent that provides the default yellow page service in the platform. The message transport system, also called agent communication channel (ACC), is the software component controlling all the exchange of messages within the platform, including messages to/from remote platforms.

JADE fully complies with this reference architecture, and when a JADE platform is launched, the AMS and DF are immediately created, and the ACC module is set to allow message communication. The agent platform can be split on several hosts. Only one Java

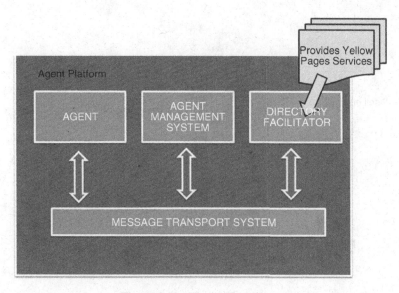

Figure 2.23 The AMS platform

application, and therefore only one Java virtual machine (JVM), is executed on each host. Each JVM is a basic container of agents that provides a complete run-time environment for agent execution and allows several agents to concurrently execute on the same host. The main container, or front-end, is the agent container where the AMS and DF reside, and where the RMI registry, that is used internally by JADE, is created. The other agent containers instead, connect to the main container and provide a complete run-time environment for the execution of any set of JADE agents. The installation of the system requires at least one computer that hosts the JADE platform. The agents may exist in this computer or other computers in communication via the internet/Ethernet (Figure 2.24).

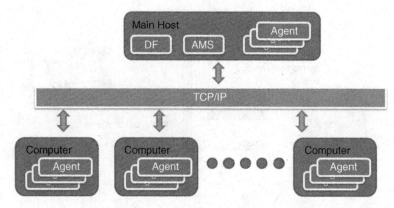

Figure 2.24 The implementation of the MAS

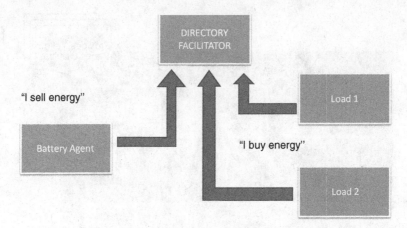

Figure 2.25 The agents declare their services to the DF agent

A critical component of this architecture is the DF, which is actually the basis for the development of plug-and-play capabilities. To further understand this, a simple example comprising a battery and two loads is considered. The procedure runs as follows: all agents, by the time they are created, automatically announce to the DF the services that they could provide to the system (Figure 2.25). In this example, the load agents participate in the system as buyers of energy, while the battery agent sells energy. The battery agent starts the transaction by sending a request to the DF. The DF agent provides the list of agents that can buy energy (Figure 2.26). Next, the battery agent sends a request to all the members of the list (Figure 2.27). Finally, the load agents respond, as shown in Figure 2.28, that is, one load agent refuses the offer, while the other agent accepts it and announces the amount of energy it needs.

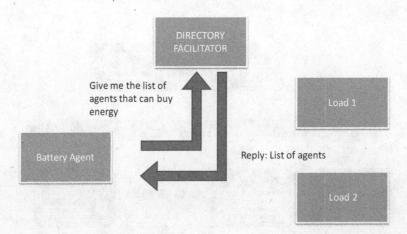

Figure 2.26 The battery agent asks for the list of agents that provide a "buying" service

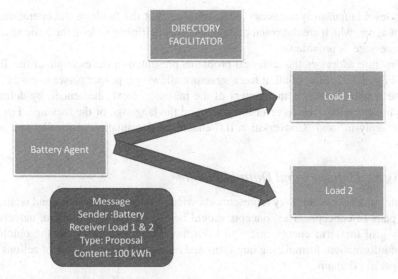

Figure 2.27 The battery agent sends message to the load agents proposing to "sell" energy

2.8.3 Agent Communication Language

The content of each message between the agents is based on the formal agent communication language (ACL) which, among others, has two main characteristics:

1. formal structure like the syntax of a human language
2. ontology which is similar to the vocabulary of the human language.

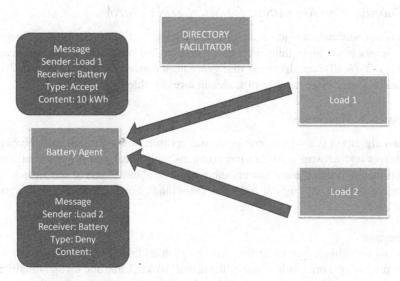

Figure 2.28 The load agents respond to the battery agent

A high-level language is necessary in order to support the fundamental operations of the intelligent agent, which are the collaboration and the intelligence. Next, the basic structure of an ACL message is provided.

This structure addresses the technical problems presented in the example of the Tower of Babel of Figure 2.22. First of all, it has a structure allowing a proper parser to easily identify who is the sender or what is the content of the message. Next, the sender, by defining the ontology it uses, allows the receiver to understand the language of the message. Finally, the attributes "reply-to" and "Conversation ID" enable the formulation of parallel and complex dialogues.

2.8.4 Agent Ontology and Data Modeling

In computer science, an ontology represents knowledge, as a set of concepts and relationships between pairs of concepts. The concepts should belong to the same domain, namely in the electricity grid the term energy refers to kWh, not calories. Agents use the ontology for passing of information, formulating questions and requesting the execution of actions related to their specific domain.

The power engineering community has devoted significant effort to defining data standards for various application areas. One example is the common information model IEC 61970 (CIM [34]) for data exchange between energy management systems and related applications. The common information model (CIM) is a unified modeling language (UML [35]) model that represents all the major objects in an electric utility enterprise typically involved in utility operations. CIM provides a set of object classes and attributes, along with their relationships. In this way, energy management system (EMS) applications developed by different vendors can exchange information. This standard cannot be directly used for the formulation of an ontology, as the agent communication language requires more complex structures than a data model. However, there is potential to use it as a basis for the development of an ontology.

2.8.5 Coordination Algorithms for Microgrid Control

In this section, coordination algorithms for decentralized control of microgrids are presented. The main issues in all algorithms are convergence to the optimal solution and scalability (complexity). Typically, an algorithm that guarantees convergence to the optimal solution cannot handle a very large number of nodes in a reasonable time.

2.8.5.1 Auction Algorithms

The auction algorithm is a type of combinatorial optimization algorithm that solves assignment problems and network optimization problems with linear and convex/nonlinear cost. The main principle is that the auctioneers submit bids to obtain goods or services. At the end of an iterative process, the highest bidder is the winner. For microgrids, goods can be an amount of energy.

English auction
A popular and very simple type of auction is the English auction. The procedure starts with the auctioneer proposing a price below that of the actual market value and then gradually raising the price. The actual value is not announced to the buyers. At each iteration, the new, higher

price is announced and the auctioneer waits to see if any buyers are willing to pay the proposed price. As soon as one buyer accepts the price, the auctioneer proceeds to a new iteration with an increased price. The auction continues until no buyers accept the new price. If the last accepted price exceeds the actual market value, the good is sold to that buyer for the agreed price. If the last accepted price is less than the actual value, the good is not sold.

Dutch auction

A similar approach is the Dutch auction. In this case, the procedure starts with the auctioneer asking a price higher than the actual value, which is decreased until a buyer is willing to accept it, or a minimum value (actual value) is reached. The winning participant pays the last announced price.

Theoretically, both approaches are equivalent and lead to the same solution. Many variations on these auction systems exist, for example, in some variations the bidding or signaling from the buyers is kept secret.

Symmetric Assignment

A more advanced auction algorithm is proposed by [36,37] to solve the symmetric assignment problem, which is formulated as follows:

Consider n persons and n objects that should be matched. There is a **benefit** a_{ij} for matching person i with object j. In the presented application, the benefit for each person is his revenues for obtaining object j, that is, an agreement for producing a certain amount of energy. The main target is to assign the persons to objects and to maximize the **total benefit,** expressed as:

$$\sum_{i=1}^{n} a_{ij} \tag{2.9}$$

The price p is an algorithmic variable that is formed by the bids of all persons and expresses the global desire. The prices of all objects form the price vector. These prices should not be confused with the market prices. Also, the difference between the benefit and the price is the actual value of an object for a specific person. The actual value for a specific object is different for two persons, since it is related to the benefit. At the beginning of the iterations, the price vector is zero and so the actual value is equal to the benefit, although variations of the proposed methods use initial non-zero values for faster convergence.

The auction algorithm calculates the price vector p, in order to satisfy the ε-complementary slackness condition suggested in [36,37]. The steps are as follows:

At the beginning of each iteration, the ε-complementary slackness condition is checked for all pairs (i,ji) of the assignment. The ji is the object j that person i wants to be assigned to. So the formulation of this condition is

$$a_{ij_i} - p_{j_i} \ge \max_{j \in A(i)} \left\{ a_{ij} - p_j \right\} - \varepsilon \tag{2.10}$$

$A(i)$ is the set of objects that can be matched with person i. This inequality has two parts: $\alpha_{ij} - p_j$ is the actual value of object j for person i, as described before. The right part refers to the object that gives maximum value to person i minus ε, where ε is a positive scalar, added in the bid of each object, in order to avoid possible infinite iterations in case two or more objects provide maximum benefit to the same person, as explained later.

If all persons are assigned to objects, the algorithm terminates. Otherwise, a non-empty subset I of persons i that are unassigned is formed. Similarly, the non-empty subset $P(j)$ is formed by the available objects. The following two steps are performed only for persons that belong to I.

The first step is the bidding phase, where each person finds an object j which provides maximal value and this is

$$j_i \in \max_{j \in A(i)} \left\{ a_{ij} - p_j \right\} \qquad (2.11)$$

Following this, the person computes a bidding increment

$$\gamma_i = u_i - w_i + \varepsilon \qquad (2.12)$$

u_i is the best object value

$$u_i = \max_{j \in A(i)} \left\{ a_{ij} - p_j \right\} \qquad (2.13)$$

and w_i the second best object value

$$w_i = \max_{j \in A(i), j \neq j_i} \left\{ a_{ij} - p_j \right\} \qquad (2.14)$$

According to these equations, the bidding increment is based on the two best objects for every person. If there are two or more bids for an object, its price rises, and the price increment is the larger bidding increment between the bids. It is obvious that, if the scalar $\varepsilon = 0$ and the benefits for the first and the second best object are the same, then $\gamma_i = 0$ and this leads the algorithm to infinite iterations. The ε scalar ensures that the minimum increment for the bids is $\gamma_i = \varepsilon$.

The next phase is the assignment phase, where each object j selected as best object by the non-empty subset $P(j)$ of persons in I, determines the highest bidder

$$i_j = \max_{i \in P(j)} \left\{ \gamma_i \right\} \qquad (2.15)$$

Object j raises its prices by the highest bidding increment $\max_{i \in P(j)} \left\{ \gamma_i \right\}$, and gets assigned to the highest bidder i_j. The person that was assigned to j at the beginning of the iteration, if any, becomes unassigned.

The algorithm iterates until all persons have an object assigned. It is proven that the algorithm converges to the optimal solution, as long as there is one. The maximum number of iterations is

$$\frac{\max_{(i,j)} \left| a_{ij} \right|}{\varepsilon} \qquad (2.16)$$

and the algorithm terminates in a finite number of iterations if

$$\varepsilon < \frac{1}{n} \qquad (2.17)$$

Application of Auction Algorithms
In order to describe how auction algorithms are applied in a MAS environment, the following simplified example is presented [38]. Two types of physical agents and one type of virtual agent are introduced. The two physical agents are the production unit agent and the load unit agent. These two agents are physical, because they directly control a production or storage unit and a load panel, respectively. The third type is the market agent. This agent is virtual because it cannot control the market in any way and just announces the prices for selling or buying energy. All other agents introduced later in this section are virtual, and their operation concerns the auction algorithm only.

Let us consider that there are x production units with a total capacity of X and y loads with a total capacity of Y. The symmetrical assignment problem requires that $X = Y$. In order to overcome the problem of surplus or deficient local production, a virtual load with a proper price is added, as shown in Figure 2.29. Similarly, virtual production can be added. The virtual load or production corresponds to the extra energy that is bought from or sold to the grid. As mentioned before, it is assumed that the grid can offer or receive infinite amounts of energy.

In order to apply the algorithm for the solution of the symmetric assignment problem, the load should be divided into equal blocks, similar to the available production. Blocks that belong to the same load have equal benefits, since the system will provide all the necessary power for the whole load or none. For example, if we consider a water heater that demands 500 Wh within the next 15 minutes, the system should provide the full 500 Wh or nothing.

Mapping the fundamental assignment problem to the microgrid management, the "persons" correspond to the blocks of available power and the "objects" to the demand blocks. The agent market operation based on the described model is illustrated in Figure 2.30. The production unit agents control the DER, the load unit agents represent the loads and the grid agent generates market player agents. The market player agents are virtual agents and their task is to accomplish the negotiation. There are two types of market player agents: the seller and the buyer. The buyer is the object in the assignment problem, and the seller is the person. Every market player agent represents a single block of energy.

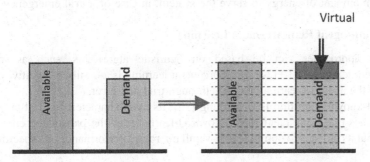

Figure 2.29 The blocks of energy that form the assignment problem. Reproduced by permission of the IEEE

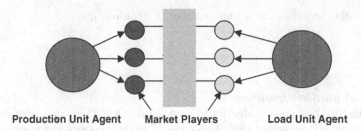

Production Unit Agent   Market Players   Load Unit Agent

Figure 2.30 The virtual market player agents that are created for the need of the negotiation. Reproduced by permission of the IEEE

Similar to the local loads, the virtual load is represented by market player agents that are created from the grid agent. According to the proposed market model, each producer has the ability to sell all its production to the grid and, similarly, every load can buy energy from the grid. For this reason, the grid agent finds the number of pairs of market player agents (sellers and buyers) and creates extra sellers and buyers. The number of the agents is equal to market agents that are created from the production unit agents and the load unit agents. In this way, buying or selling energy from the grid is determined by the algorithm.

A major issue in microgrid operation is the estimation of the upper limits for the demand or the available power of each DER for the next time interval. This should be done separately for each participant. It should be noted that although forecasting techniques are well advanced for larger interconnected systems and typically hourly resolution times, there is little experience in forecasting with a high temporal resolution (e.g. <15 minutes) with a horizon of 3–4 hours for very small loads, like the loads in a microgrid, as already discussed in Section 2.6.

In this application the upper limit is defined by two methods, depending on the type of load or DER. The first method is to consider that the upper limit is the nominal capacity. For units such as a diesel generator or a water heater this is quite realistic. By contrast, for units like photovoltaic panels, wind generators or lighting loads, the persistence method is used, i.e. it is assumed that the average energy production or demand for the next 15 minutes will be the same as the current one.

It should be noted that other functionalities of the microgrid (such as security check, battery management and voltage control) can be included in this operation. For example, the offered power of battery bank production could be reduced in order to maintain the state of charge and keep certain amount of energy to serve the system, in case of a grid emergency.

2.8.5.2 Multi-agent Reinforcement Learning

Alternative approaches can be based on heuristic algorithms, such as multi-agent reinforcement learning (RL) [22]. Reinforcement learning is a family of iterative algorithms that allows the agent to learn a behavior through trial and error.

One well-known algorithm is Q-learning [39]. Its main characteristic, in the multi-agent environment is that each agent runs its own Q-learning for the part of the environment it perceives, but its target is to optimize the overall microgrid performance or a specific common goal.

Q-learning is a reinforcement learning algorithm that does not need a model of its environment and can be used online. The Q-learning algorithm operates by estimating the values of state-action pairs. The value $Q(s,a)$ is defined as the expected discounted sum of future payoffs obtained by taking action α from state s and following an optimal policy thereafter. Once these values have been learned, the optimal action from any state is the one with the highest Q-value. After being initialized, Q-values are estimated on the basis of experience, as follows:

- From the current state s, select an action a. This will bring an immediate payoff r, and will lead to a next state s'.
- Update $Q(s,a)$ based upon this experience as follows: $Q(s, a) = (1 - k)Q(s, a) + k(r + \gamma \max_{a'} Q(s', a'))$ where k is the learning rate, $0 < \gamma < 1$ is the discount factor.

Alternative learning algorithms can be used, including the Nash-Q learning, which is a general sum MAS reinforcement algorithm for a stochastic environment. The main problem is that the execution times are prohibitive because the Nash-Q learning requires that the Q table includes as a parameter the actions of the other agents. For systems like microgrids, the Q table becomes huge, requesting a large number of episodes for training. Other approaches propose forecasting the decisions of other agents, but this also cannot be easily done in microgrid applications.

The main problem in the application of reinforcement learning is the modeling of the environments which affects the size of the Q table and as a consequence the convergence speed. All actions and system states are included in the Q table as

$$Q(s, \alpha1, \alpha2, \alpha3, \ldots \alpha n) \tag{2.18}$$

$a1, a2 \ldots an$ is the selected action of agent 1, agent 2 $\ldots$ agent n.

Another concern is that the environment is stochastic, since we still cannot predict accurately the effect on the system state after switching a load or sending a command to change the setpoint of a DG. An approach proposed in [40,41] is to replace all actions with one single variable called transition, which represents the final result to the environment of all actions of all the agents.

$$\underbrace{Q(s, \alpha1, \alpha2, \alpha3, \ldots \alpha n)}$$

$$Q(s, \alpha1, tr) \tag{2.19}$$

The agent selects the action that will lead the system to the best state (transition) considering that the other agents will follow the same policy. The selection of actions is based on the following equation:

$$\text{Selected transition} = \arg\max_{tr=1,2,3} \left(\sum \max_{\alpha}(Q(s, a, tr)) \right) \tag{2.20}$$

This means that, for each transition, each agent selects the action that maximizes the Q value and they all add these values at each transition. The selected transition is the one with the highest value.

Let's consider an example of the application of the RL algorithm to microgrid black start. After the black out, a simple procedure is followed:

1. switch off all loads
2. launch black start units
3. launch the other units
4. start the MAS according to the results of the reinforcement learning

It should be noted that the algorithm considers the steady state of the system and does not handle transient phenomena. The algorithm focuses on how to ensure power supply to the critical loads for a predefined period, for example 24 hours ahead. In this case, the agents have to learn to use the available resources in the most efficient way.

Each agent executes a Q-learning procedure for the part of the environment that it perceives. For the formulation of the problem, the variables that will be inserted in the Q table should be defined first. It should be noted again that it is important to keep the size of the Q table small in order to reduce the number of calculations.

The environment state variable forms a table of 24 elements, one for every hour of the schedule. The production units are characterized by one more variable, the available *fuel*, with three values: {low, medium, high}. For battery units this variable reflects the *state of charge*. Finally, for the loads, there is a variable called *remain* with values {low, medium, high}, indicating how many hours they need to be served.

The *transition* variable is considered next, with three values: {up, neutral, down}. This variable is an indication of the behavior of the other agents and the state of the system, as explained before. The purpose of this variable is to identify the most likely next states of the system. For example, if the *transition* has value {down}, the system will go to a worse state, no matter what the action of the individual agent might be. The definition of a worse or better state depends on the type of problem. For example, in the interconnected system operation a worse case is when the system receives energy from the upstream networks. In non-interconnected mode, the worse state is when the system consumes stored energy.

Accordingly, the size of the Q table for each agent is:

- storage units: Q (horizon{24}, fuel{3}, environment {3}, transition {3},action {3}) = 1944 elements
- generation units: Q (horizon{24}, fuel{3}, environment {3}, transition {3},action {2}) = 1296 elements
- loads: Q (horizon{24}, environment {3}, remain{3}, transition {3}, action {2}) = 1296 elements.

The agent learns the value of its action in the various states of the system. For this case study, the agents are able to act as in Table 2.7.

The intermediate reward for the algorithm is calculated from:

$$\text{reward} = N * (\text{transition reward} + \text{final state reward} + K) \qquad (2.21)$$

N is a normalization parameter obtained by dividing the maximum production capacity of the unit by the total production capacity of the system. This ensures a weighted

Table 2.7 Actions of the agents. Reproduced by permission of the IEEE

	Type	Actions
1	load	on
		off
2	storage unit	produce
		stop
		store
3	production unit	stop
		produce

participation of all units, for example, a 100 kW unit affects more the final actions than a 1 kW unit.

- The transition reward has a value of 1/24 when the system stores energy, 0 if it remains at the same level and −1/24 when the system consumes stored energy.
- The final state reward is received in the final step and has value 1 if the system has sufficient energy for the whole period (24 h), and −1 if not.
- The K parameter is different for loads and production/storage units. For production/storage units it indicates the remaining fuel, and for load the time (in percentage that the load should be served).

This algorithm needs to be executed if there is a significant change in the system, such as the installation of a new unit. This is presented in Figure 2.31. After the execution, every agent has learnt what to do in case of an emergency. Consider for example, that the system is in zero power exchange with the slack. The agents have to select one of three transitions {up, neutral, down}. In order to decide which transition to follow, they announce to each other the Q values for each transition in the current state. The selected transition is given by Eq. (2.20).

Selecting for example an "up" transition means that some agents have surplus power and they offer it to the system having in mind that the selected path will lead to a good final solution. The good solution is the one that ensures energy adequacy for the whole 24 hour period.

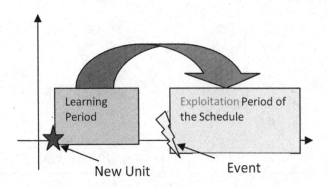

Figure 2.31 Time schedule of the algorithm. Reproduced by permission of the IEEE

As an example, consider a microgrid system comprising:

- diesel generator,
- battery bank,
- load,
- renewable energy sources.

The simulation has two parts. The first part is training (exploration), in order to find the Q values. The second part is exploitation of the algorithm in isolated operation. Several simulations of the operation of the system were performed in order to validate whether the agents find the solution that ensures energy adequacy. Furthermore, simple software was developed, allowing each agent to decide absolutely independently of each other, in order to compare the solution with the one of the reinforcement learning algorithms.

The critical loads and the renewable energy sources participate in the simulation of the exploitation, but there is no need to train the respective agents, since they do not control their actions.

A learning rate $k = 0.95$ and discount factor $\gamma = 0.1$ are assumed. The algorithm converges after 20 000 iterations, which means that there are no significant changes in the values of the Q table in more iterations. In order to ensure that this is the final solution, multiple runs have been made with the same schedule, but with different initializations of the Q table, as well as multiple runs with the same initial Q table. Since there is no interaction between the agents during the learning period, every agent needs around 40 seconds in a single PC with 3 GHz processor to complete the training.

In Figure 2.32, an instant of the results of the algorithm for the battery is shown. The vertical axis presents the Q value and the horizontal axis the time step. The agent chooses the action with the higher Q value. The battery agent appears to learn how to handle the islanded operation. The agent starts exhibiting a conservative behavior at the beginning, since it does not know what will happen next, so the system tries to save energy for the next hours. This is shown by the fact that the battery Q values are higher for zero production in the first 8 hours, in

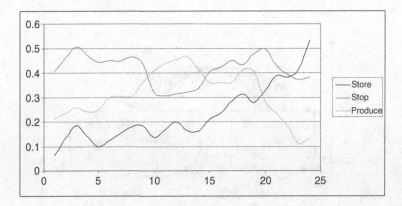

Figure 2.32 Results for restoration study case

comparison to the Q values for production or storing. By the time the energy adequacy is guaranteed, the agents start to serve extra loads, like the battery, in the hours between 10 and 15. It should be noted that this is a simplified example. In a more complex application, the Q values for the battery should be compared with the corresponding Q values of the other agents.

2.8.6 Game Theory and Market Based Algorithms

Another interesting approach is to use game theory and market based rules. The various agents know their own benefit and cost, and they respond to price signals. This approach is using the principles of game theory, and if the rules of the game are correct, the system will balance at the optimal point.

Game theory is a branch [42,43] of applied mathematics that studies the interaction of multiple players in competitive situations. Its goal is the determination of the equilibrium state at which the optimal gain for each individual is achieved. More specifically, the theory of non-cooperative games studies the behavior of agents in any situation where each agent's optimal choice may depend on its forecast of the choices of its opponents [43]. Various categories of games exist, depending on the assumptions regarding the timing of the game, the knowledge associated with the payoff functions and last but not least the knowledge regarding the sequence of the previously made choices. More specifically, the games can be categorized as:

- static/dynamic games: the players choose actions either simultaneously or consecutively.
- complete/incomplete information: each player's payoff function is common knowledge among all the players or at least one player is uncertain about another player's payoff function.
- perfect/imperfect information (defined only for dynamic games): at each move in the game the player with the move knows or does not know the full history of the game thus far [42].

A simple but realistic example is to assume a dynamic game of complete and perfect information: the consumers and the DG units are not considered as competitive entities, but their payoff functions are publicly available, while the history of the game at each stage is known. The timing of such a game is as follows:

1. Player 1 chooses an action ($a1$) from a feasible set of actions ($A1$).
2. Player 2 observes this action and then chooses an action ($a2$) from its feasible set ($A2$).
3. Payoffs are $u1$ ($a1$, $a2$) and $u2(a1, a2)$.

The solution of such a game is determined as the backwards-induction outcome. At the second stage of the game, player 2 will solve the following problem, given the action $a1$ previously chosen by player 1:

$$\max_{a_2 \in A_2} u_2(a_1, a_2) \tag{2.22}$$

It is assumed that for each $a1$ in $A1$, player 2's optimization problem has a unique solution, denoted by $R2(a1)$ [42]. This is player 2's best response to player 1's action. Since player 1 can solve player 2's problem as well as player 2 can, player 1 should anticipate player 2's reaction to each action $a1$ that player 1 might take, so player 1's problem at the first stage amounts to

$$\max_{a_1 \in A_1} u_2(a_1, R_2(a_1)) \tag{2.23}$$

It is assumed that this optimization problem for player 1 also has a unique solution, denoted by a_1^*. So $(a_1^*, R_2(a_1^*))$ is the backwards induction outcome of the game.

A detailed description of game theory concepts can be found in [42,43].

2.8.7 Scalability and Advanced Architecture

A key question about the applicability of decentralized approaches and especially of MAS systems, concerns their suitability for control of larger systems with several hundreds of

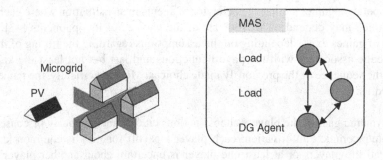

Case 1 Single Microgrid

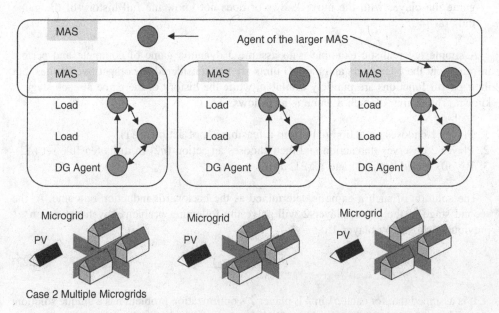

Case 2 Multiple Microgrids

Figure 2.33 General scheme of MAS architecture

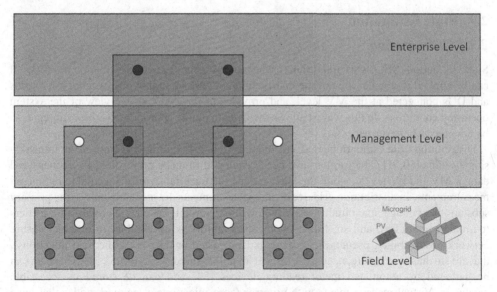

Figure 2.34 Management of the agents

nodes, as real microgrids can be. This problem is referred to as the scalability problem, and it is very important for the development of any control system. It should be noted that service oriented architectures and cloud computing provide technical tools to address this issue, but these are beyond the scope of this book and we will focus in the main concept in this section.

This concept can be better explained by considering the organization of human societies and cities. Many people living in an area form a village or a city. Next, many cities and villages form a county and many counties form a country. This concept is shown in Figure 2.33. Accordingly, DG units and controllable loads form small microgrids and accordingly small MASs. These MASs form larger MASs and so on. The grouping may be realized based on electrical and topological characteristics such as having a common MV transformer.

The groups of MASs are organized in three levels. The three levels are presented in Figure 2.34. All the agents associated directly with the control of the production units or controllable loads belong to the field level. These agents directly communicate and control a production unit or a load and may be organized in a MAS according to the physical constraints of the system. Each of these MASs also has an agent that is responsible for communicating with other higher-level MASs, in order to cooperate with them. These MASs belong to the management level. Finally, these MASs may form larger MASs in order to participate at the enterprise level.

From a microgrid point of view, the field level of Figure 2.34 is associated with each individual microgrid control, the management level is associated with multi-microgrids and individual DGs at MV, as further discussed in Chapter 5, while the enterprise level is related to a higher level aggregation, such as coordinated market participation.

2.9   State Estimation

2.9.1   Introduction

State estimation (SE) is very important for the management of active distribution networks. It can be applied to a wider area of the distribution network including one or more microgrids and DGs connected at the MV level, and provides to the DSO an overview of the system operating conditions. In this way, it allows the DSO to define appropriate control strategies to be adopted, whenever necessary.

Distribution state estimation (DSE) techniques [44–49] are different from SE in transmission systems [50,51]. The former have been developed to make up for the lack of measured data at MV and LV levels, while the latter reduce the uncertainty of the available redundant measurements. In distribution grids, real-time measurements are available only at the primary substations (voltage magnitudes, power flows and circuit breaker statuses) and feeders (current magnitudes), and so, full network observability is impossible. In order to ensure network observability, pseudo-measurements (forecasted or near real-time load injections), that are stochastic in nature, need to be used at all unmeasured nodes [45,52,53]. This data can be gathered by automated meter reading devices [54] and can be stored in accessible databases. Virtual measurements with no error (zero injections at network nodes that have neither load nor generation, zero voltage drops at closed switching devices and zero power flows at open switching devices) can be also utilized. The DSE will process this real-time and forecasted data to produce the state vector consisting of nodal voltages (magnitudes and phase angles) and transformer tap positions.

Feeders are mainly three-phase radial, but have laterals that can be single- or two-phase. Furthermore, loads on the feeders are distributed and can be single- and two-phase (for residential service) or three-phase (for commercial and industrial service). Therefore, distribution systems are unbalanced by nature. Nevertheless, to avoid modeling complexities, the network is assumed to be balanced, and the single phase equivalent network model is considered for state estimation analysis.

The commonly used weighted least-squares (WLS) estimation method [51] can be adopted for distribution state estimation, considering the following nonlinear measurement model:

$$z = h(x) + e \tag{2.24}$$

The nodal states can be estimated by minimizing the quadratic objective function:

$$\min_{x} J(x) = (z - h(x))^{T} R^{-1} (z - h(x)) \tag{2.25}$$

where z is the measurement vector, $h(x)$ is the vector of nonlinear functions relating measurements to states, x is the true state vector, e is the vector of normally distributed measurement errors, with $E(e) = 0$ and $E(ee^{T}) = R = diag(\sigma_i^2)$, and σ_i^2 is the variance of the ith measurement error. Real-time measurements will have lower variance than pseudo-measurements. The state estimate $\hat{x}$ can be obtained by iteratively solving the following equations:

$$G(x^k)\Delta x^k = H^T(x^k)R^{-1}\Delta z^k \tag{2.26}$$

where, k is the iteration index, $\Delta x^k = x^{k+1} - x^k$, $\Delta z^k = z - h(x^k)$, $H = \frac{\partial h}{\partial x}$ is the Jacobian matrix and $G(x^k) = H^T(x^k)R^{-1}H(x^k)$ is the gain matrix.

The presence of a large number of load injection pseudo-measurements may give rise to convergence problems [55]. In order to overcome this problem, robust SE algorithms need to be applied [50]. One such algorithm, based on orthogonal transformations, is presented in Section 2.9.2.

With the increasing number of nodes involved in distribution networks, the DSE may not be suitable for operating as a centralized algorithm. The feeders can be divided into zones or areas, and local state estimations can be executed independently, sending their results to the DMS, where a coordinated state estimator will calculate the system-wide state [56,57]. When partial lack of communication occurs in some areas, the local SE processes can continue with the remaining areas.

2.9.2 Microgrid State Estimation

The structure of a typical LV distribution network, including a microgrid system, connected to the main MV grid, has been shown in Figure 2.4. A microgrid state estimator (MSE) will follow the concepts of distribution state estimators, receiving a limited number of real-time measurements from the network [47,58]. Near-real-time measurements of voltage and active and reactive injections at DG sites can be also available at predefined time intervals. Since this data is inadequate for state estimation, forecasted node injections obtained from historical or near-real-time load data should be used. However, the number of loads connected to the distribution network may be large, so it will be impractical to telemeter all those points.

Measurement time skew is a consideration when combining large area data received via a data communication network. Due to limited communication infrastructure, near real-time data from DG sources or loads may be sent to the DMS through different communication channels causing additional time skew problems. In order to accommodate the effects of randomly varying arrival of measurement data, a stochastic extended Kalman filter (EKF) algorithm can be used to improve time skewed measurement data [59].

An important issue in state estimation modeling is the identification of network configuration (topology). In transmission state estimation the statuses of switching devices are processed by the network topology processor (NTP) to define the bus/branch network model, by merging bus sections joined by closed switching devices into nodes. The topology is assumed to be known and correct, but any status errors, that pass undetected by the NTP, will result in an incorrect bus/branch model. In distribution networks including DGs, it is frequently not possible to find and fix one topology with a very high certainty, due to frequent topology changes (switching operations for network reconfiguration, in/out of service load, branch or DG or microgrid system islanding). This issue is discussed further in Chapter 4 (see, for example, Figure 4.8). In any case, one topology must be considered to initiate the SE process, but the formulation should be flexible enough to allow changes in the topology, if the initial one will not lead to the best solution. This means that the NTP must be able to consider that the identification of bad data can find errors in the status of some switching devices. This problem can be solved by augmenting the conventional state vector with switching device statuses and other related pseudo-measurements in order to identify topology changes and errors [51,60–62], that is, the topology will be estimated at the same time with analog information.

In distribution systems with several microgrids, an additional problem is that the number of islands is not known when starting the state estimation process, as a consequence of some switching devices having an unknown or suspicious status. A network splitting problem can be formulated as the problem of finding the state variables in all network islands. The consideration of uncertainty affecting topology introduces the splitting problem. When the network is split into two or more non-connected electrical islands – owing to a set of switching devices being reported as open – the system becomes unobservable, and the state vector cannot be computed. In order to overcome this problem, a degree of uncertainty should be considered for the pseudo-measurements. By including such pseudo-measurements, the network becomes observable, and the state vector can be estimated [60].

In order to set up the measurement system, we assume that $\mu_i = z_i$ is the mean value of the ith measurement. Then a $\pm 3\sigma_i$ deviation around the mean covers about 99.7% of the Gaussian curve. Hence, for a given percentage of maximum measurement error about the mean μ_i, the standard deviation σ_i is given by [44]

$$\sigma_i = \frac{\mu_i \times \text{error}\%}{300} \tag{2.27}$$

In practice, an error of 1% for voltage measurements, 3% for power flow and injection measurements and 15% for load pseudo-measurements may be considered.

Since the load pseudo-measurements are statistical in nature, we use the state error covariance matrix as a performance index to assess the accuracy of the state estimation solution [50]:

$$C_x = \left(H^T R^{-1} H\right)^{-1} \tag{2.28}$$

The ith diagonal entry $C_x(i, i)$ of C_x is the variance of the ith state.

Simulations studies [44] show that the effect of the inaccuracies of the load estimates are more severe at the unmeasured load buses, because of the low local measurement redundancy. As expected, there is a strong correlation between bus voltage uncertainties and the errors in load estimates. It should be noted that, uncertainties for all buses adjacent to the primary substation are low, because of the higher accuracy and redundancy of the local real-time measurements. This fact enables the correct identification of topology errors and network splitting. From state estimation runs, it can be shown that an error in a load value on the upper feeder in network of Figure 2.4 affects mainly the node voltages of the feeder where this load belongs to and not those of the lower feeder [44]. When applied to real microgrids, the state estimate accuracy mostly depends on the accuracy of the load models.

2.9.3 Fuzzy State Estimation

Usually, the SE problem is solved by using all the information available for the network, and not just measurement values. Evidently, the quality of the solution depends critically on the quality of the available information. In order to take this into account, a different model using fuzzy state estimation (FSE) can be also applied [63,64], using information characterized by uncertainty applying fuzzy set theory. Fuzzy numbers are used to model

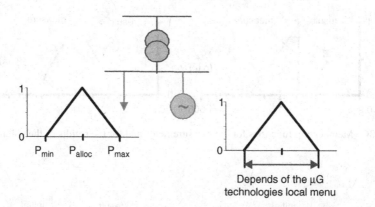

Figure 2.35 Type of fuzzy information considered for the microgrid load and generation

this kind of information, and these are used as model input data (can be called fuzzy measurements). One source of fuzzy measurements is obtained from some "typical" load curve that defines a band of possible values for the load, based, for example, on a historical database. In this way, it is possible to define a fuzzy assessment for the actual active load value. If the microgrid generation is not measured, a procedure to define fuzzy measurements can be used based on the mix of type of technology and all useful forecasted values available.

An FSE algorithm that exploits fuzzy measurements and involves qualitative information about the type of load and generation can incorporate both deterministic traditional measurements obtained by measurement devices – even if affected by metering errors – and fuzzy measurements obtained by fuzzy evaluations or resulting from load allocation procedures [65].

The FSE algorithm uses in the first phase a crisp measurement vector to run a crisp weighted least squares SE algorithm to compute the state vector. In the second phase, the fuzzy deviations specified for the measurements are reflected in the results of the SE [1]. Active and reactive power flows, currents in lines and transformers, power injected by generators or by connections with other networks, active and reactive load values are computed using fuzzy algebra [66].

Even in cases when all communications with microgrids are missing, qualitative data can be used to replace the missing measurements. This qualitative data corresponds to fuzzy measurements/variables, an example of which is shown in Figure 2.35.

The FSE algorithm is executed under these conditions to produce an estimation of bus voltages (Figure 2.36). Other results obtained from the FSE algorithm are the values for the power injections at each bus (Figure 2.37), power flows and current on each network branch.

FSE allows the integration of qualitative data for the loads (when they are not measured or communications are missing) and the integration of qualitative data for DG, and these uncertainties are reflected in the results. For instance, the value of the load at the bus corresponds to the symmetric boundaries of the values of power injection. These active and reactive power injections are included in the input data as fuzzy measurements. On these membership functions the central values of the initially specified and computed membership functions are slightly shifted. This is understandable, considering that this set of values is used

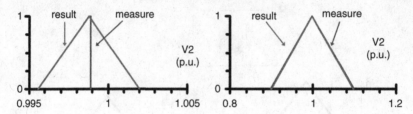

Figure 2.36 Membership functions for the measurements and for the results of the voltage magnitude

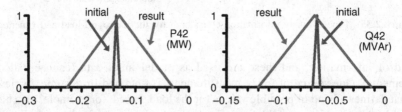

Figure 2.37 Membership functions for the measurements and for the results of the active and reactive power injection

to perform the initial crisp MSE study aimed at obtaining a coherent operating point for the system. This means that, due to metering errors and to fuzzy assessments, this set of input values does not correspond to a coherent picture, possibly not being in accordance with Kirchhoff's laws, and therefore the errors are filtered by the MSE procedure. The results have one central value that corresponds to the most likely value for the load, but the load can be in the neighborhood of this value (described by the resulting membership function).

2.10 Conclusions

This chapter provides a framework for microgrid energy management. An overview of the microgrid control architectures and their main functionalities are discussed. The basic distinction between centralized and decentralized approaches is highlighted, identifying the benefits and characteristics of each approach. Centralized functionalities are formulated, and the results from their indicative application in a typical LV microgrid, adopting different policies in market conditions, are presented. Special focus is placed on intelligent decentralized control using multi-agent (MAS) technologies. The basic features of intelligent agents are provided, including practical implementation issues. Discussion about the forecasting needs and expectations and state estimation requirements at distribution level are also included.

Appendix 2.A Study Case Microgrid

The resistances and reactances of the lines of the study case network of Figure 2.15 are shown in Table 2.8. The values are expressed per unit on a power base of 100 kVA and voltage base 400 V.

Table 2.8 R and X of the lines of the study case network

Sending	Receiving	R (pu)	X (pu)
Grid	1	0.0025	0.01
1	2	0.0001	0.0001
2	3	0.0125	0.00 375
3	4	0.0125	0.00 375
4	5	0.0125	0.00 375
5	6	0.0125	0.00 375
3	7	0.021 875	0.004 375
1	8	0.033 125	0.00 875
1	9	0.0075	0.005
9	10	0.015	0.010 625
10	11	0.02 125	0.005 625
11	12	0.02 125	0.005 625
9	13	0.010 625	0.005 625
13	14	0.010 625	0.005 625
10	15	0.023 125	0.00 625
15	16	0.023 125	0.00 625

Table 2.9 Installed DG sources

Unit ID	Unit type	Min power (kW)	Max power (kW)
1	MT	6	30
2	FC	3	30
3	WT	0	15
4	PV1	0	3
5	PV2	0	2.5
6	PV3	0	2.5
7	PV4	0	2.5
8	PV5	0	2.5

Table 2.9 provides the capacity of the DG sources and summarizes their bids. The efficiency of the fuel consuming units and the depreciation time for their installation have been taken into account, as shown in Table 2.10. The term c_i denotes the payback compensation of the investment for each hour of operation for fuel consuming units. For renewable energy sources, the investment payback corresponds to term b_i. For simplicity, term a_i is assumed to be zero.

The depreciation time and values for installation cost are summarized in Table 2.11. In all cases, the interest rate is 8%. Both micro-turbine and fuel cell are assumed to run on natural gas whose efficiency is 8.8 kWh/m^3 [11]. For the microturbine the efficiency is assumed to be 26% for burning natural gas, while the efficiency of a fuel cell is assumed to be 40% [12]. Data from [12,13,16] has been used for the life time of the DG and the installation cost, as summarized in Table 2.11. Using Eq. (2.29) the cost per year can be calculated for every type of DG. This cost is distributed either to the operating hours of the DG sources that consume fuel or to the production of the intermittent DG units such as wind turbines or PV. For MT and FC, we have assumed that they operate for 90% of the year or 7884 hours. For WT, we have

Table 2.10 Bids of the DG sources

Unit type	a_i ($\in$ct/kWh^2)	b_i ($\in$ct/kWh)	c_i ($\in$ct/h)
MT	0	4.37	85.06
FC	0	2.84	255.18
WT	0	10.63	0
PV1	0	54.84	0
PV2	0	54.84	0
PV3	0	54.84	0
PV4	0	54.84	0
PV5	0	54.84	0

Table 2.11 Financial data for determining the bids

	MT	FC	WT	PV
Life-time (years)	12.5	12.5	12.5	20
Costs in bibliography[a] ($\in$/kW)	800–2000	3000–20 000	800–5000	4200–10 000
Installation cost ($\in$/kW)	1500	4500	2500	7000
Depreciation time (years)	10	10	10	20
Depreciation cost ($\in$/kW-year)	223.54	670.62	372.57	712.92

[a] These costs do not necessarily reflect current costs, especially because capital costs of some technologies have been significantly reduced.

assumed 40% capacity factor which means 3504 kWh/kW and for the PVs the yearly production is 1300 kWh/kW according to [16].

$$\text{Ann\_Cost} = \frac{i(1+i)^n}{(1+i)^n - 1} \cdot \text{InsCost} \qquad (2.29)$$

i is the interest rate, n the depreciation time in years, InsCost, the installation cost and Ann_Cost is the annual cost for depreciation.

References

1. More Microgrids. [Online] www.microgrids.eu.
2. Lasseter, R., Akhil, A., Marnay, C. *et al.* (2002) White Paper on Integration of Distributed Energy Resources. The CERTS Microgrid Concept. CA: Tech. Rep. LBNL-50829, Consortium for Electric Reliability Technology Solutions (CERTS).
3. Katiraei, F. *et al.* (2008) Microgrids management. IEEE Power and Energy Magazine, 6, 54–65.
4. W3C. [Online] http://www.w3.org/.
5. Charytoniuk, W. and Chan, M.S. (2000) Short-term load forecasting using Artificial Neural Networks. A review and evaluation. *IEEE T. Power Syst.*, **15** (1), 263–268.
6. Steinherz Hippert, H., Pedreira, C.E. and Souza, C.S. (2001) Neural Networks for short-term load forecasting, a review and evaluation. *IEEE T. Power Syst.*, **16** (1), 44–55.
7. Taylor, J. (2008) An evaluation of methods for very short-term electricity demand forecasting using minute-by-minute British data. *Int. J. Forecasting*, **24**, 645–658.
8. Anemos project. [Online] http://www.anemos-project.eu.

9. Giebel, G., Kariniotakis, G. and Brownsword, R., The State-of-the-art in short-term prediction of wind power. A literature overview. Deliverable Report D1.1 of the Anemos project (ENK5-CT-2002-00665), available online at http://anemos.cma.fr.

10. Liu, K. *et al.* (1996) Comparison of very short-term load forecasting techniques. *IEEE T. Power Syst.*, **11** (2), 877–882.

11. Canu, S., Duran, M. and Ding, X. (1994) District heating forecast using artificial neural networks. *Int. J. Eng.*, **2** (4).

12. Paravan, D., Brand, H. *et al.* (2002) Optimization of CHP plants in a liberalized power system. Proceedings of the Balkan Power Conference, vol 2, pp. 219–226.

13. Nogales, F.J. *et al.* (2002) Forecasting next-day electricity prices by time series models. *IEEE T. Power Syst.*, **17** (2), 342–348.

14. García-Martos, C., Rodríguez, J. and Sánchez, M.J. (2011) Forecasting electricity prices and their volatilities using unobserved components. *Energ. Econ.*, **33**, 1227–1239.

15. Sideratos, G. and Hatziargyriou, N. (2012) Probabilistic wind power forecasting using radial basis function neural networks. *IEEE T. Power Syst.*, **27**, 1788–1796.

16. Pinson, P. and Kariniotakis, G. (2004) Uncertainty and Prediction Risk Assessment of Short-term Wind Power Forecasts. Delft, The Netherlands: The Science of Making Torque from Wind.

17. Tsikalakis, A.G. and Hatziargyriou, N.D. (2008) Centralized control for optimizing microgrids operation. *IEEE T. Energy Conver.*, **23** (1), 241–248.

18. Hatziargyriou, N.D., Dimeas, A. and Tsikalakis, A. (2005) Centralised and decentralized control of microgrids. *Int. J. Distr. Energ. Resour.*, **1** (3), 197–212.

19. Tsikalakis, A.G. and Hatziargyriou, N.D. (2007) Environmental benefits of distributed generation with and without emissions trading. *J. Energ. Policy*, **35** (6), 3395–3409.

20. Rao, S.S. (1996) Chapter 7, in *Engineering Optimization,Theory and Practice*, John Wiley & Sons, New York.

21. Papadogiannis, K.A., Hatziargyriou, N.D. and Saraiva, J.T. (2003) *Short Term Active/Reactive Operation Planning in Market Environment using Simulated Annealing*, ISAP, Lemnos, Greece.

22. Lee, K.Y. and El-Sharkawi, M.A. (2002) *Tutorial Modern Heuristic Optimization Techniques with Applications to·Power Systems*, IEEE PES., Chicago.

23. Papathanassiou, S., Hatziargyriou, N. and Strunz, K. (2005) A Benchmark LV microgrid for Steady State and Transient Analysis. Cigre Symposium "Power Systems with Dispersed Generation", Athens,Greece..

24. Kariniotakis, G.N., Stavrakakis, G.S. and Nogaret, E.F. (1996) Wind power forecasting using advanced neural networks models. *IEEE T. Energy Conver.*, **11** (4), 762–767.

25. APX. Amsterdam Power Exchange. [Online] http://www.apx.nl.

26. McArthur, S.D.J. *et al.* (2007) Multi-agent systems for power engineering applications— Part I: concepts, approaches, and technical challenges. *IEEE T. Power Syst.*, **22** (4), 1743–1752.

27. McArthur, S.D.J. *et al.* (2007) Multi-agent systems for power engineering applications—Part II: technologies, standards, and tools for building multi-agent systems. *IEEE T. Power Syst.*, **22** (4), 1753–1759.

28. Ferber, J. (1999) *Multi-Agent Systems. An introduction to Distributed Intelligence*, Addison-Wesley.

29. Bradshaw, J.M. (1997) *Software Agents*, MIT Press.

30. Wooldridge, M. (2009) *An Introduction to Multi-agent Systems*, 2nd edn, John Wiley & Sons.

31. Dimeas, A.L., Hatzivasiliadis, S.I. and Hatziargyriou, N.D. (2009) Control agents for enabling customer-driven microgrids. IEEE, Power & Energy Society General Meeting, PES'09, IEEE.

32. JADE website. [Online] www.jade.tilab.com.

33. FIPA website. [Online] www.fipa.org.

34. CIM. IEC 61970. [Online] http://www.iec.ch/smartgrid/standards/.

35. UML. [Online] http://www.uml.org.

36. Bertsekas, D.P. (1992) Auction algorithms for network flow problems: a tutorial introduction. *Comput. Optim. Appl.*, **1**, 7–66.

37. Castanon, D.P. and Bertsekas, D.A. (1992) A forward/reverse auction algorithm for asymmetric assignments problems. *Comput. Optim. Appl.*, **1**, 277–297.

38. Dimeas, A. and Hatziargyriou, N.D. (2005) Operation of a multi-agent system for microgrid control. *IEEE T. Power Syst.*, **20** (3), 1447–1455.

39. Dayan, P. and Watkins, C.J. (1992) Q-learning. *Mach. Learn.*, **8**, 279–292.

40. Veloso, M. and Peter, S. (1999) Opaque-Transition Reinforcement Learning. Proceedings of the Third International Conference on Autonomous Agents.

41. Dimeas, A.L. and Hatziargyriou, N.D. (2010) Multi-agent reinforcement learning for microgrids. IEEE, Power and Energy Society General Meeting.

42. Gibbons, R. (1992) *Game Theory for Applied Economists*, Princeton University Press, Princeton, New Jersey.
43. Fudenberg, D. and Tirole, J. (1991) *Game Theory*, MIT Press, Cambridge, Massachusetts.
44. Korres, G.N., Hatziargyriou, N.D. and Katsikas, P.J. (2011) State estimation in multi-microgrids. *Euro. Trans. Electr. Power Special Issue: Microgrids and Energy Management*, **21** (2), 1178–1199.
45. Ghosh, A.K., Lubkeman, D.L. and Jones, R.H. (1997) Load modeling for distribution circuit state estimation. *IEEE T. Power Deliver.*, **12** (2), 999–1005.
46. Kelley, A.W. and Baran, M.E. (1994) State estimation for real-time monitoring of distribution systems. *IEEE T. Power Syst.*, **9** (3), 1601–1609.
47. Thornley, V., Jenkins, N. and White, S. (2005) State estimation applied to active distribution networks with minimal measurements. 15th Power Systems Computation Conference, Liege, Belgium: August 2005.
48. Ghosh, A.K., Lubkeman, D.L., Downey, M.J. and Jones, R.H. (1997) Distribution circuit state estimation using a probabilistic approach. *IEEE T. Power Syst.*, **12** (1), 45–51.
49. Singh, R., Pal, B.C. and Jabr, R.A. (2009) Choice of estimator for distribution system state estimation. *IET Gener. Transm. Distrib*, **3** (7), 666–678.
50. Exposito, A.G. and Abur, A. (2004) *Power System State Estimation: Theory and Implementation*, Marcel Dekker, New York.
51. Monticelli, A. (1999) *State Estimation in Electric Power Systems: A Generalized Approach*, Kluwer Academic Publishers, Boston, US.
52. Wang, H. and Schulz, N.N. (2006) Using AMR data for load estimation for distribution system analysis. *Electr. Pow. Syst. Res.*, **76**, 336–342.
53. Wang, H. and Schulz, N.N. (2001) A load modelling algorithm for distribution system state estimation. Conference and Exposition in Transmission and Distribution, vol. 1 102–105.
54. Samarakoon, K., Wu, J., Ekanayake, J. and Jenkins, N. (2011) Use of Delayed Smart Meter Measurements for Distribution State Estimation. IEEE PES General Meeting, July 2011.
55. Gu, J.W., Clements, K.A., Krumpholz, G.R. and Davis, P.W. (1983) The solution of ill-conditioned power system state estimation problems. *IEEE T. Power Ap. Syst.*, **PAS-102**, 3473–3480.
56. Korres, G.N. (2011) A distributed multiarea state estimation. *IEEE T. Power Syst.*, **26**, 73–84.
57. Gomez-Exposito, A., de laVilla Jaen, A., Gomez-Quiles, C. *et al.* (2011) A taxonomy of multi-area state estimation methods. *Electr. Power Syst.*, 81 1060–1069.
58. Cobelo, I., Shafiu, A., Jenkins, N. and Strbac, G. (2007) State estimation of networks with distributed generation. *Eur. Trans. Electr. Power*, **17**, 21–36.
59. Su, C.-L. and Lu, C.-N. (2001) Interconnected network state estimation using randomly delayed measurements. *IEEE T. Power Syst.*, **16** (4), 870–878.
60. Korres, G.N. and Manousakis, N.M. (2012) A state estimation algorithm for monitoring topology changes in distribution systems. PES General Meeting, San Diego, CA, USA, 2012, pp. 1–7.
61. Pereira, J. (May 2009) A state estimation approach for distribution networks considering uncertainties and switching. Porto: Faculdade de Engenharia da Universidade do Porto. PhD Thesis, July 2001 [Online] http://saca.inescporto.pt/artigos/artigo150.pdf.
62. Katsikas, G.N. and Korres and, P.J. (2002) Identification of circuit breaker statuses in WLS state estimator. *IEEE T. Power Syst.*, **17** (3), 818–825.
63. Saric, A.T. and Ciric, R.M. (2003) Integrated fuzzy state estimation and load flow analysis in distribution networks. *IEEE Trans. Power Delivery*, **18** (2), 571–578.
64. Konjic, T., Miranda, V. and Kapetanovic, I. (2005) Fuzzy inference systems applied to LV substation load estimation. *IEEE T. Power Syst.*, **20** (2), 742–749.
65. Miranda, V., Pereira, J. and Saraiva, J.T. (2000) Load allocation in DMS with a fuzzy state estimator. *IEEE T. Power Syst.*, **15** (2), 529–534.
66. Zadeh, L.A. (1965) Fuzzy Sets. *Inform. Control*, **8**, 338–353.

3

Intelligent Local Controllers

Thomas Degner, Nikos Soultanis, Alfred Engler and Asier Gil de Muro

3.1 Introduction

The development and use of the intelligent local controllers for microgenerators and controllable loads enables stable microgrid operation. These controllers provide efficient voltage and frequency control in case of islanded operation, ancillary services in interconnected mode, and deal efficiently with frequency and voltage variations during transitions from interconnected to islanded operation and vice versa.

The main tasks of microsource controllers are to allow power sharing between the different sources at different locations, acting on the electronic converters that interface them to the network. As communication lines, especially for long distances and control purposes are expensive and vulnerable, at least fast communication should be avoided. This can be achieved by using local controllers, which are based on frequency and voltage droops. This way, the locally available information – frequency and voltage – is used for power sharing. Normally, the frequency is linked to the active power, and voltage is linked to the reactive power.

Standard rotating generator systems inherently support these droops. Sometimes, for improving performance, additional controllers are applied. Therefore it is possible to operate large numbers of such generators in a single network. However, with standard static inverters, characterized by fixed frequency and voltage, such an approach is not possible. A control approach providing frequency and voltage droops is required.

In this chapter, intelligent control concepts for power electronics interfaced microsources are described. Control strategies for multiple inverter systems are developed with reference to the self-synchronizing mechanism of conventional rotating generators. The appropriate control of frequency and voltage in each inverter facilitates the sharing of the active and reactive power in the system. Simulation results that demonstrate power sharing between distributed generators are given. In the following, the implications of line parameters, in particular of the high resistance practically found in LV microgrids, are systematically

Microgrids: Architectures and Control, First Edition. Edited by Nikos Hatziargyriou.
© 2014 John Wiley & Sons, Ltd. Published 2014 by John Wiley & Sons, Ltd.
Companion Website: www.wiley.com/go/hatziargyriou_microgrids

discussed. Finally, a control algorithm using the fictitious impedance method is analyzed and characteristic simulation results are provided.

3.2   Inverter Control Issues in the Formation of Microgrids

3.2.1   Active Power Control

The basic principle that allows classic synchronously rotating generators to change their power output in response to a change in the system load without an explicit communication network is the frequency variation at the machine terminal as a function of power demand. When two points in the network are operating at different frequencies there is an increase of active power delivery from the location of higher frequency to the location of lower frequency. As this happens, the two frequencies tend to drift towards a common average value until the new steady state is reached (self-synchronizing torque). This physical relation between active power output and frequency in the synchronous generators has to be implemented in the control of the inverters, in the case of a system with inverter interfaced sources.

3.2.2   Voltage Regulation

Voltage control must ensure that there are no large circulating reactive currents between sources. Without local voltage control, systems with high penetration of microsources could experience voltage and/or reactive power oscillations. The issues are identical to those involved in the control of large synchronous generators. However, in large power systems the impedance between generators is usually large enough to impede circulating currents. In a microgrid, that is typically radial, the problem of large circulating reactive currents is immense. With small errors in voltage setpoints, the circulating current can exceed the ratings of microsources. This situation requires a voltage vs. reactive current droop controller so that, as the reactive current generated by the microsource becomes more capacitive, the local voltage setpoint is reduced. Conversely, as the current becomes more inductive, the voltage setpoint should increase.

3.3   Control Strategies for Multiple Inverters

3.3.1   Master Slave Control Scheme

The control method that has been used for paralleling inverter units in the case of UPS systems is very difficult to adopt for a distributed generation application like the microgrid. The load current may be measured and divided among the participating inverter units, but this requires that the load and the microsources are separated.

In the master–slave control, one inverter regulates the voltage and the resulting current of this unit defines the current references of the other units. The master acts as a voltage source inverter and the rest of the units as current source inverters [1].

A system with a voltage source, as master and additional controllable current sources (grid supporting units) is depicted in Figure 3.1. The supervisory control is responsible for power distribution. The main features of this approach can be summarized as follows:

- a simple control algorithm at component level is required
- high expenditure for buses and their cabling is needed

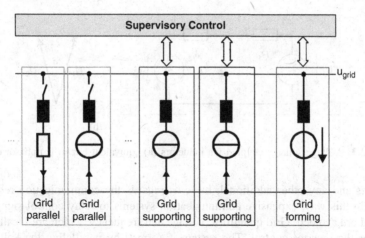

Figure 3.1 Principle of a (classic) modular supply system consisting of one voltage source, current sources and passive loads

- the system is difficult to expand
- supervisory control is required

As a result, this approach is not suitable for the operation of systems based on distributed generation, mainly due to high communication requirements and the needs for supervisory control and extra cabling, even for small systems.

3.3.2 Multi-Master Control Scheme

Communication and/or extra cabling can be avoided, if the inverters themselves set their instantaneous active and reactive power. In [2] a concept has been developed that uses reactive power/voltage and active power/frequency droops for the power control of the inverters (Figure 3.2). Droops function in a similar way to those used by conventional rotating machines in large systems, as discussed in Section 3.2. The supervisory control just provides parameter settings for each component, which may comprise the idle frequency f_0 and voltage u_0, the

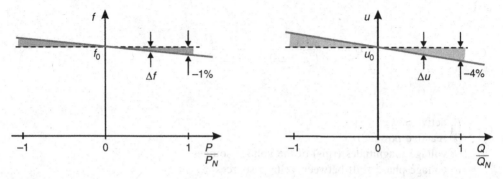

Figure 3.2 Frequency and voltage droops. Reproduced by permission of Technology & Science Publishers [9]

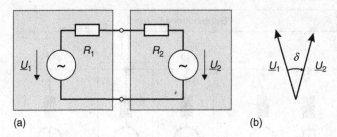

(a)   (b)

Figure 3.3   Voltage sources coupled via inductors (a) equivalent circuit (b) phasor diagram

droop slopes and any other additional, basic as regards the communication requirements, command. In this way, expensive communication systems to relay control signals to each inverter unit are avoided. Grid quantities, voltage and frequency, which are locally available are used for the inverter control. The system is formed by paralleling the voltage source inverters (VSIs) that interface the microsources via their filters and the network impedances. The benefits of this approach are the following:

- simple expansion of the system
- increased redundancy with VSI paralleling. Master/slave operation is abandoned and the system does not rely on vulnerable communications.
- optimization based on a simple bus system.
- a simplified supervisory control with more complex control tasks assigned to the components.

Two inverters in parallel are shown in Figure 3.3, considering only the output voltages at the fundamental frequency. The inverters are coupled via the inductances resulting from cabling and harmonic filters. The active power P and the reactive power Q of the voltage sources coupled by inductors with negligible resistance can be calculated as follows:

$$P = \frac{U_1 \cdot U_2}{\omega \cdot (L_1 + L_2)} \cdot \sin(\delta) \tag{3.1}$$

and

$$Q = \frac{U_1^2}{\omega \cdot (L_1 + L_2)} - \frac{U_1 \cdot U_2}{\omega \cdot (L_1 + L_2)} \cdot \cos(\delta) \tag{3.2}$$

P: active power
Q: reactive power
U_1, U_2: voltage magnitudes (rms) of the voltage sources
δ: voltage phase shift between voltage sources
ω: cycle frequency of the grid
L_1, L_2: coupling inductances

A phase shift δ between two voltage sources causes active power flow, while reactive power flow is due to voltage difference $U_1 - U_2$. Assuming standard values for the inductances L_1, L_2, results in very sensitive systems, where even small deviations of the voltage phase and magnitude cause high current flows between the inverters. This sensitivity is the reason why fixed frequency and voltage inverter concepts fail. There is always a voltage difference due to tolerances of the sensor, references, temperature drift and aging (e.g. 1–5%). Also, crystals used for the time reference are not identical, and the angle difference is the integral of the frequency error over time. Therefore, precise control with complex algorithms is required for the parallel operation of voltage source inverters.

A precise active power dispatch is required for balancing demand and generation. Applying droops, two approaches are possible: $P(f)$ and $f(P)$ that is, measuring frequency and defining active power output or vice versa.

3.3.3 Droop Control Implementation in the VSI

The direct control system analogy of the droop line, as applied to synchronous machines, is to measure system frequency and control real power (Figure 3.4). System frequency is measured with a phase-locked loop (PLL) technique, which operates based on the three-phase terminal voltage. The system frequency is compared with a reference value (typically 50 Hz under normal operation). The frequency deviation is filtered using a low-pass filter and multiplied by a gain constant to obtain the droop control [3].

For mainly inductive power systems, that is, systems with a high ratio of reactance to resistance (X/R), Eq. (3.2) expresses the fact that differences in voltage cause reactive power flows, or conversely, reactive power flows influence terminal voltages. Typically therefore, reactive power is controlled by a Q vs. V droop, as shown in Figure 3.2 [3].

Figure 3.5 shows the block diagram of the voltage regulation control technique. Three-phase terminal voltages (V_{ta}, V_{tb}, V_{tc}) are measured and fed as inputs to the controller. The magnitude of the terminal voltage vector (V_{tmag}) is calculated and compared with the set reference value (e.g. $V_{tref} = 415$ V). The voltage error is filtered by a low-pass filter and multiplied by a gain constant to obtain the droop control of the VSC. The output of the voltage regulation control block provides the reactive power (Q_{injV}) that needs to be injected to maintain the terminal voltage according to the droop set value.

However, it is not straightforward to obtain accurate measurement of instantaneous frequency in a real system. Measuring instantaneous real power is easier. It has therefore been proposed [4–6] that control is implemented in the opposite way, that is, the output power

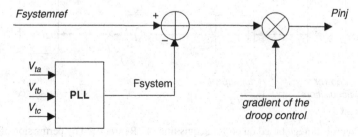

Figure 3.4 Block diagram of the VSC control for frequency regulation

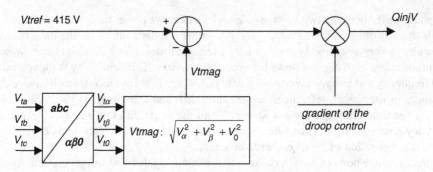

Figure 3.5 Block diagram of the VSC control for voltage regulation

of the voltage source converter is measured and this quantity is used to adjust the output frequency, as shown in Figure 3.6. Similarly, reactive power is measured and used to adjust the voltage magnitude. Active and reactive power are estimated from the inverter voltage and output current in the power acquisition block. A first-order delay is used for passing through each of the P, Q values. These may be part of the power measurement filters and serve to decouple the two control channels and to adjust the time frame within which the inverter frequency and voltage change with respect to power changes. The droop controls are applied next, to define the voltage reference of the inverter. Active power determines the frequency and reactive power determines the voltage magnitude through the respective droops. It is also noted that a phase correction is introduced in direct proportion to the measured active power. This branch enhances the system stability.

Figure 3.7 shows power sharing among three SMA Sunny Island™ single phase inverters programmed with this scheme (rated power 3.3 kW, clock 16 kHz, coupling inductor 0.8 mH). They are connected to a resistive load, each with an LV cable of approximately 10 m. The frequency droop of the inverters denoted by L_1, L_2 was set to 1 Hz/rated power, while the

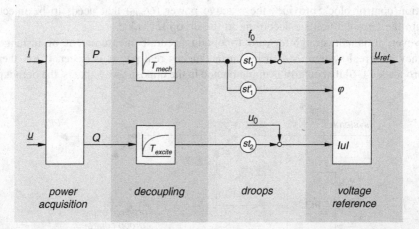

Figure 3.6 Control strategy based on power acquisition [4]. Reproduced by permission of Technology & Science Publishers [9]

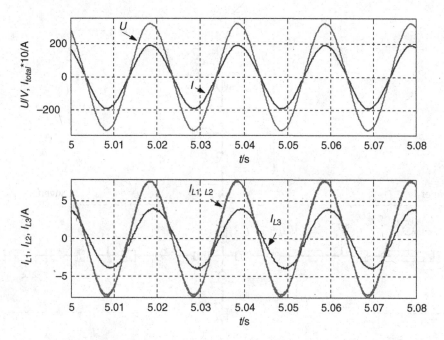

Figure 3.7 Parallel operation of three identical inverters sharing a resistive load via their frequency vs. power droop control. Reproduced by permission of Technology & Science Publishers [9]

droop of the inverter denoted by L_3 was set to 2 Hz/rated power. It is evident that this method allows the L_3 inverter to supply half the power of the other two and that active power sharing is defined by the droop constants. It is remarkable that reactive power is circulating – as seen from the inverter current – as a result of the phase difference caused between inverter L_3 and L_1, L_2.

Next, a simulated case of a microgrid is presented. The microgrid (Figure 3.8), consisting of three inverters coupled via an MV distribution system (15 kV), is simulated in ATP/EMTP (alternative transients program/electromagnetic transients program: http://www.emtp.org/). The inverters are represented by controllable three-phase voltage sources, and the distribution system consists of switches, overhead lines (π-blocks), transformers and the load. The three inverters operate in parallel via the MV distribution system and supply a resistive load with a total power of 100 kW. The inverters are controlled with the algorithm described in Figure 3.6.

With suitable setting of the slope of the frequency droop the contribution of the inverters is 20, 30 and 50 kW, as seen in Figure 3.9. The slope of the frequency droop may be used in order to take into account the size of the inverter. The idle frequency f_0 can be used to control the energy flow. The idle frequency of the inverters in this case is set to 50 Hz. Owing to the load, the system frequency decreases to 49 Hz (Figure 3.10). The frequency could be restored to 50 Hz with the use of a secondary control.

The setting of the slope of the voltage droop results in a variable virtual inductance between the respective inverter and the distribution system. It should be chosen in order to ensure stable operation. The idle voltage should be set in order to minimize reactive power. However, voltage limits have to be respected.

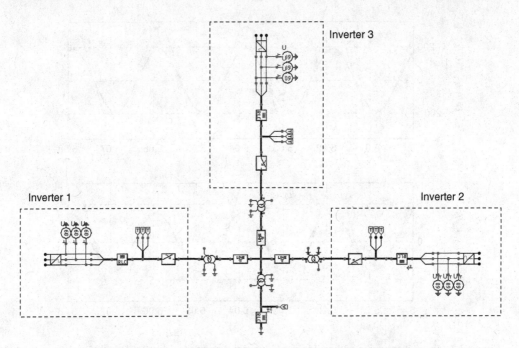

Figure 3.8 Simulation of inverter dominated system in ATP/EMTP

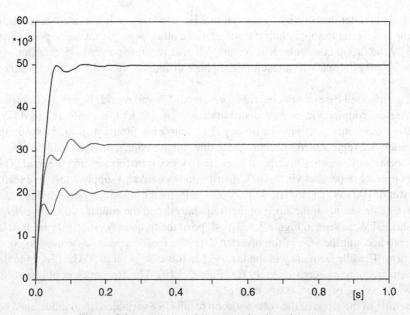

Figure 3.9 Load shared between the three inverters

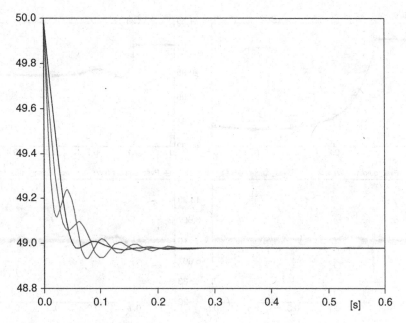

Figure 3.10 Frequency change in the islanded microgrid

3.3.4 Ancillary Services

While in grid-connected mode, microgrids can contribute to the overall grid control and thus facilitate the operation of the system by providing ancillary services.

A microgrid can support voltage in the area of its connection to the upstream network, in the event of a disturbance that causes a voltage drop. Figure 3.11 shows the simulation results for the sudden reduction of voltage at the point of common coupling (PCC) of a microgrid that is connected at the end of a long supply feeder with and without voltage regulation. The microgrid with voltage regulation, in this case the VSC of an energy storage unit, injects reactive power to maintain the rms terminal voltage (V_{trms}) within acceptable limits.

The central controller effectively sets the droop line slope of the individual VSC interfaces to determine reactive power sharing between the units and the reactive power export/import of the microgrid under grid-connected operation. During islanded operation the net reactive power flow will be zero. The voltage in the microgrid will then increase (if there is excess reactive power), forcing power electronics interfaced units to produce less Q or even absorb reactive power, until a new steady-state voltage is reached (net zero Q flow).

Here the voltage/reactive power droops act in order to limit the reactive power flow. The voltage control and the flow of reactive power in LV grids, which are mainly resistive, are discussed in the following paragraphs.

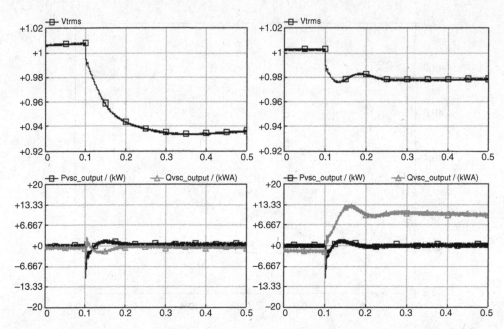

Figure 3.11 Simulation results without (a) and with (b) voltage regulation control of the microgrid connected to main supply

A microgrid can absorb/inject a specified amount of aggregate power at the point of common coupling. This is carried out by adjusting the position of the droop lines of the individual VSCs from a central controller. A simulated case is shown in Figure 3.12. This adjustment does not require a particularly fast control loop, since even a slow telecommunications link has time constants significantly faster than most power network subsystems. The additional active and reactive powers ($Pinj$, $Qinj$) to be injected in this case are set by the central controller for each specific VSC. Alternatively, the droop line settings could have been adjusted for each VSC by the central controller directly. In the system simulated, two limit blocks are used to limit the power injection (active power limits ± 10 kW, reactive power limits ± 20 kVAr). These limits are important during the transient period caused by a disturbance in the system.

3.3.5 Optional Secondary Control Loops

If a microsource has to produce a specific active power during the isolated operation, a secondary control action may be employed. The control forces the active power output of the VSI to remain at P_{ref} while the system frequency changes, by acting on the f_{ref} value. The droop line is shifted vertically as shown in Figure 3.13.

The microgrid frequency will end up in steady state after every load change to a value different from nominal that depends on the droop slope and the relation between the load and frequency. To restore the frequency to the nominal value, a secondary control loop is needed to cancel out the residual difference. The control action changes the P_{ref} value so that the

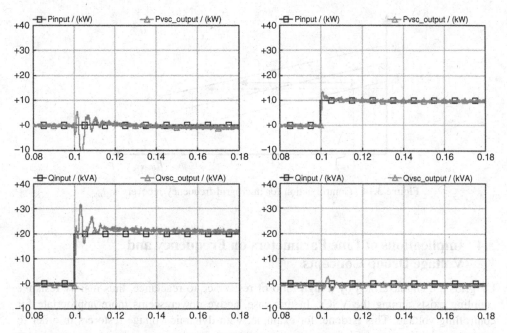

Figure 3.12 Simulation results with the active and reactive power control set by central controller (all other controls are deactivated and the microgrid is connected to the main supply)

frequency is always brought back to the f_{ref} value. The droop line is shifted horizontally as shown in Figure 3.14.

Likewise, secondary control action can be applied to the voltage/reactive power control. Reactive power output of the VSI may be kept to a constant value Q_{ref} by changing the V_{ref} value and shifting the droop line vertically as in Figure 3.13 for active power. To keep the output voltage of the inverter constant, Q_{ref} may be changed, shifting the droop line horizontally as is done in Figure 3.14 for frequency.

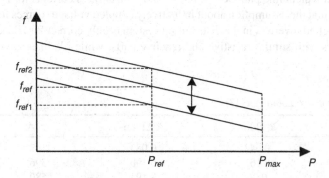

Figure 3.13 Control action for constant active power production at P_{ref}

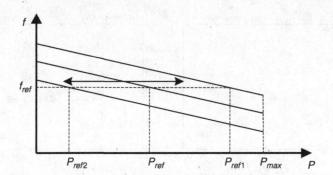

Figure 3.14 Control action for microgrid frequency recovery at f_{ref}

3.4 Implications of Line Parameters on Frequency and Voltage Droop Concepts

LV grids are characterized by a high ratio of resistance to reactance, thus mainly resistive coupling exists among the VSCs. In this case, active power seems more appropriate for controlling voltage. This is done, for example, with dynamic voltage restorers in order to compensate voltage sags (short-term interference).

The use of active power flow for voltage control would have financial consequences – it would change completely the meaning of economic active power dispatch. Also the compatibility with rotating generators in the LV grid and the upwards compatibility with the main grid would be lost.

The function of voltage/reactive power droops in the LV grid is to limit reactive power flows. Keeping the voltage in the allowed range should be done by a suitable layout of the LV lines and eventually taking care to increase the reactive part of the lines with the addition of chokes.

In this section, we analyze the effect of the network line parameters in the application of the droop control of the VSI units that form the microgrid.

3.4.1 Power Transmission in the LV Grid

Table 3.1 shows typical line parameters, R', X', and typical rated currents for HV, MV and LV lines. It is clear that the assumption about inductively coupled voltage sources for representing the droop controlled inverters in the distribution system is only correct for HV lines. MV lines have parameters with similar resistive and reactive parts, while LV lines are even predominantly resistive.

Table 3.1 Typical line parameters [7]

Type of line	R' Ω/km	X' Ω/km	IN A	R'/X'
LV line	0.642	0.083	142	7.7
MV line	0.161	0.190	396	0.85
HV line	0.06	0.191	580	0.31

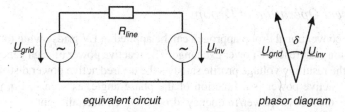

equivalent circuit · phasor diagram

Figure 3.15 Resistively coupled voltage sources. Reproduced by permission of Technology & Science Publishers [9]

The active power P_{inv} and the reactive power Q_{inv} of resistively coupled voltage sources – here an inverter and a grid – can be calculated as follows, with the notation adopted in Figure 3.15:

$$Q_{inv} = \frac{U_{inv} \cdot U_{grid}}{R_{line}} \sin \delta \tag{3.3}$$

$$P_{inv} = \frac{U_{inv}^2}{R_{line}} - \frac{U_{inv} \cdot U_{grid}}{R_{line}} \cos \delta \tag{3.4}$$

Equation (3.4) shows that, in LV grids active power flows are coupled with voltage, while phase difference between the voltage sources causes reactive power flows (Eq. (3.3)). This fact suggests using active power/voltage and reactive power/frequency droops – hereinafter called "opposite droops" – in the LV grid, instead of reactive power/voltage and active power/ frequency droops – "conventional droops".

3.4.2 Comparison of Droop Concepts at the LV Level

In the following, the advantages and disadvantages of using conventional or opposite droops at the LV level are summarized in Table 3.2.

As seen in Table 3.2, the only advantage of using opposite droops is the direct voltage control. But if the voltage was controlled in this way, no power dispatch would be possible. Each load would be fully supplied by the nearest generator. As this is generally not possible, voltage deviations would remain in the grid. Using conventional droops results in connectivity at the HV level, allows power sharing with rotating generators and provides a precise power dispatch. The voltage deviations within the grid depend on the grid layout, which is today's standard.

Table 3.2 Comparison of droop concepts for the LV level

	Conventional droop	Opposite droop
Compatible with HV-level	yes	no
Compatible with rotating generators	yes	no
Direct voltage control	no	yes
Active power dispatch	yes	no

3.4.3 Indirect Operation of Droops

Basically, the conventional droop approach can be applied in LV grids owing to the generator voltage variation with reactive power exchange. The reactive power of each generator is tuned in a way that the resulting voltage profile satisfies the desired active power distribution. In the LV grid, the reactive power is a function of the phase angle, as seen in Eq. (3.3). This is adjusted with the active power/frequency droop. Of course, the entire loop has to be consistent. Four stable operating points depending on the slopes of the droops can be distinguished, two of which make sense [8,9].

3.4.3.1 Stable Operating Points

In order to derive the operating points of conventional droops in LV grids, the stability of the power transfer Eq. (3.4) is assessed in a simplified manner. Since at a reasonable operating point, δ is rather small, $\cos \delta$ becomes almost 1. Rearranging (3.4) with given power P_{inv} and given grid voltage U_{grid}, two solutions for the inverter voltage U_{inv} are obtained from the quadratic equation:

$$U_{inv1,2} = \frac{U_{grid}}{2} \pm \sqrt{\frac{U_{grid}^2}{4} + P_{inv} \cdot R} \tag{3.5}$$

Both solutions are considered next. U_{inv1} is a voltage close to the grid voltage and U_{inv2} is a slightly negative voltage. This implies a 180° phase shift. Therefore, in (3.6) the factor k is introduced, with $k_1 = 1$ for the first solution and $k_2 = -1$ for the negative solution. The factor k is an approximation of the cos function in (3.4):

$$P_{inv} = \frac{U_{inv1,2} - U_{grid}}{R} = U_{inv1,2} \cdot k_{1,2} \tag{3.6}$$

The inverter power P_{inv} is adjusted by changing the inverter voltage U_{inv} with its reactive power, as follows:

$$U_{inv1,2} - U_{grid} = U_{inv1,2} \cdot q_{droop}, \tag{3.7}$$

From Eq. (3.3) Q_{inv} is a function of the angle δ, for small values of which

$$U_{inv1,2} \approx \delta \cdot \frac{U_{inv1,2} \cdot U_{grid}}{R} \tag{3.8}$$

δ is the integral over time of the generator's frequency difference from the grid frequency:

$$\delta = \int \Delta f \, dt \tag{3.9}$$

and

$$\Delta f = (P_{set} - P_{inv}) \cdot p_{droop} \tag{3.10}$$

Table 3.3 Stable operating points of conventional droops in LV grids

Case	Description	p_{droop}	q_{droop}	k	Comment
1	inverse conv.	pos.	pos.	1	acceptable
2	conv.	neg.	neg.	1	acceptable
3		pos.	neg.	-1	not acceptable
4		neg.	pos.	-1	not acceptable

The integral character of this process ensures the above-mentioned precise power distribution. With backward substitution from (3.10) to (3.6) and solving for P_{inv}:

$$P_{inv} = \int P_{set} - P_{inv} dt \cdot \underbrace{\frac{p_{droop} \cdot q_{droop} \cdot k_{1,2} \cdot U_{grid} \cdot U_{inv1,2}^2}{R^2}}_{=C},$$ (3.11)

which describes a first-order lag equation with solution:

$$P_{inv} = P_{set}(1 - e^{-C \cdot t}).$$ (3.12)

This simplified solution is obtained by the assumptions that U_{inv} is constant with time and that the power acquisition's dynamic is neglected. Equation (3.12) is stable, only if the constant C is positive, which means that

$$p_{droop} \cdot q_{droop} \cdot k_{1,2} > 0,$$ (3.13)

with p_{droop} and q_{droop} as the respective droop factors. The four stable operating points derived from (3.13) are summarized in Table 3.3.

Cases 1 and 2 (the inverse conventional and the conventional droop) are characterized by the same sign of both droop factors. This requires k to be 1, which results in an inverter voltage which is near the grid voltage (3.5). Only little reactive power is needed to tune the voltage, whereas in cases 3 and 4, huge reactive power is needed. Even worse, the inverter power and a huge amount of grid power are dissipated in the line. Therefore, cases 3 and 4 are unacceptable.

3.4.3.2 Simulation

In this paragraph, the inverter voltage U_{inv}, the grid voltage U_{grid} and the voltage drop across the line U_{line} are simulated for all four cases of Table 3.3. A single phase system is modeled assuming a line resistance of $0.5\,\Omega$. The injected inverter power is $10\,kW$. The results were obtained using the simulation tool Simplorer^{TM}.

1. Case 1 (Figure 3.16a) : In principle, this operates satisfactorily and 3.3 kVar reactive power is needed for tuning the inverter's voltage. The current (in phase with U_{line}) lags the grid voltage U_{grid}. Thus, the inverter is loaded inductively, and according to the reactive power/ voltage droop its voltage is increased. However, the droops of this case would not work

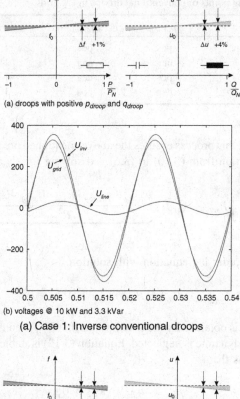

(a) Case 1: Inverse conventional droops

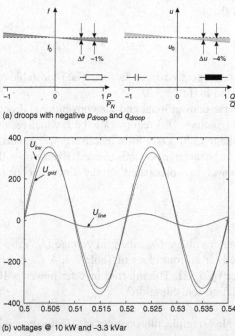

(b) Case 2: Conventional droops

Figure 3.16 Comparison of droop concepts. Reproduced by permission of Technology & Science Publishers [9]

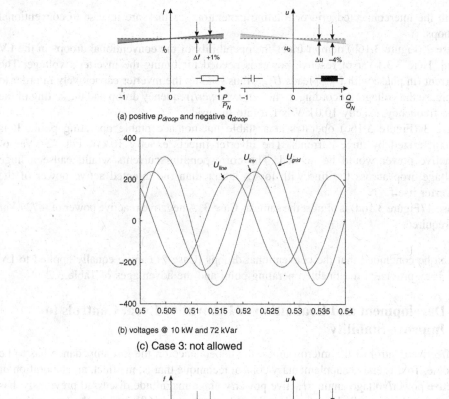

(a) positive p_{droop} and negative q_{droop}

(b) voltages @ 10 kW and 72 kVar

(c) Case 3: not allowed

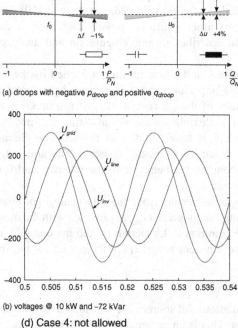

(a) droops with negative p_{droop} and positive q_{droop}

(b) voltages @ 10 kW and -72 kVar

(d) Case 4: not allowed

Figure 3.16 (*Continued*)

with the interconnected grid or rotating generators, as they are inverse to conventional droops.

2. Case 2 (Figure 3.16b) demonstrates the operability of the conventional droops in the LV grid. Here 3.3 kVar of reactive power is needed for tuning the inverter's voltage. The current (in phase with U_{line}) leads U_{grid}, thus loading the inverter capacitively in order to increase the voltage. According to the active power/frequency droop and the setting of the idle frequency, exactly 10.0 kW is fed into the grid.

3. Case 3 (Figure 3.16c) operates at a stable but not acceptable operating point. It is characterized by huge currents. The inverter injects exactly 10 kW, but 72 kVar of reactive power would be needed. The corresponding currents would cause a huge voltage drop across the line with losses larger than the injected active power of the inverter itself.

4. Case 4 (Figure 3.16d) is almost the same as Case 3. A negative reactive power of -72 kVar is required.

It can be concluded that the conventional droops (Case 2) can be equally applied to LV grids. They provide a reasonable operating point and the advantages of Table 3.2.

3.5 Development and Evaluation of Innovative Local Controls to Improve Stability

For effective control of the microsources, the dependence on the line impedance has to be overcome. To this end, a supplementary control technique that is, in effect, an elaboration of the active power/voltage angle, reactive power/voltage magnitude discussed previously, has been proposed; it is called the fictitious impedance method [10]. It uses the instantaneous current feedback, thus providing additional possibilities such as control of the harmonic frequencies. This section describes its implementation and analyzes the results by using simulations.

A voltage drop is imposed on the inverter output formed by the inverter current across a fictitious impedance. This impedance value is a control parameter and is defined at will, according to the conditions of the network or the microsource/inverter stage of operation. Therefore, it constitutes a flexible means of adjusting the inverter to the system. The effectiveness of the control is tested under all formed line characteristics that may be encountered, from fully inductive to mostly resistive. The aim is to decouple the P/f and Q/V control channels, changing the behavior of the inverter, and leading to a system with inductively coupled sources.

In the following paragraphs, results from stability analysis of distribution networks with resistive and inductive coupling impedances are provided with the main emphasis on resistive coupling, which is more common in LV networks and the one that provokes more stability problems. The alternative transient program (ATP) is used for the simulations.

3.5.1 Control Algorithm

In order to simplify simulations, all sources are considered as fully controllable, three-phase balanced voltage sources. This is an acceptable simplification, since the inverter allows the control of the output voltage waveform.

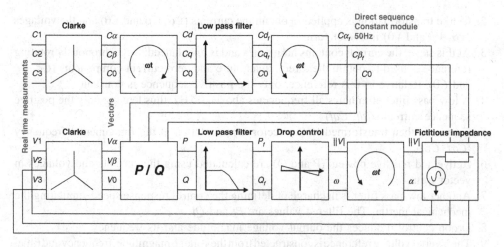

Figure 3.17 Control of voltage and frequency

The output impedance of a source facilitates its parallel connection with other sources, and the formation of the system. In the case of the microsource inverter, this impedance is assumed to be the coupling transformer leakage inductance or the inverter filter reactance. In addition to this output impedance, the line impedance must also be taken into consideration. In the LV network this is mainly resistive but it is modeled as a series connection of resistance and reactance to take into account all possibilities.

As discussed, the strong coupling of real and reactive power in LV affects the droop control techniques. In order to solve this problem, a fictitious impedance is introduced to change the behavior of the coupling between inverters and to improve the system stability. An outline of the control is shown in Figure 3.17. In order to avoid harmonics and voltage unbalances, the output current is filtered first and the resulting positive sequence is multiplied by the impedance and then added to the output voltage.

In summary, the following steps are involved:

- current is measured and filtered
- active and reactive power are calculated using the output voltage
- P and Q power is filtered to smooth the response
- droop controls are applied
- output fictitious impedance is simulated

The fictitious impedance constitutes an additional control variable. It can be defined arbitrarily in terms of magnitude and angle. It can be reactive, resistive or of any ratio in between. It is noted though that, contrary to a physical resistance, the fictitious resistance implies no losses, and the efficiency of the inverter is not compromised [11].

In detail, the algorithm steps are as follows:

1. Currents ($C1$, $C2$ and $C3$) are measured from the real-time sampled values and voltages ($V1$, $V2$ and $V3$) are taken from their calculated output values.

2. Clarke transformation is applied to obtain the currents ($C\alpha$, $C\beta$ and $C0$) and the voltages ($V\alpha$, $V\beta$ and $V0$) in vector form.
3. At this stage, the current contains harmonics and is unbalanced: a synchronously rotating reference is used to obtain the direct, quadrature and zero current components (Cd, Cq and $C0$). Relative to this reference, only the positive sequence is constant.
4. A low-pass filter eliminates all frequencies above 100 Hz, thus leaving only the positive sequence current (Cdf, Cqf)
5. Current is then transformed to the vector representation at the fundamental frequency ($C\alpha f$, $C\beta f$).
6. Active and reactive powers (P and Q) are calculated using the currents and voltages in vector form.
7. Another low-pass filter is in charge of defining the control response speed, emulating the mechanical inertia. The filtered values are Pf and Qf.
8. Droop control produces the output voltage magnitude and its frequency.
9. The inverter voltage reference is constructed from the voltage magnitude, frequency and time.
10. Finally the fictitious impedance technique is included, adding a voltage drop across this impedance to the final output voltage.

It is remarkable that the Q/V droop can be eliminated, because the effect of the reactive power flowing through the fictitious impedance is a voltage drop, with the same slope direction as the Q/V control droop. Therefore, both can be considered as complementary control strategies or, in order to simplify the control algorithm, an equivalent voltage drop may substitute for the Q/V droop result.

The reaction time and the frequency response of the digital filters operate as the inertia of the conventional rotating mass generators. If the response is too slow, the device will react as a heavy rotor machine. The response will be slow and the high-frequency variations of power will be negligible. On the other hand, if the time response of the digital filter is fast, the device will appear as a low inertia power source and any variation of the output power will affect its frequency. If the power consumption in the microgrid changes considerably, the devices with bigger inertia (e.g. with slower filters) will absorb most of the power fluctuations.

The magnitude of the fictitious impedance is usually chosen higher than that of the filter impedance so that it dominates the total output inverter impedance. An inductive fictitious impedance may alter the picture in a predominantly resistive LV network and enhance system stability, avoiding the necessity to increase the filter inductance. The resistive part can reduce the high frequency oscillations that appear with a predominantly inductive coupling as will be discussed. The resistive component could also be employed to share unbalanced currents and harmonics.

Choosing the fictitious impedance inversely proportional to the rated power of the inverter would automatically lead to current sharing of the inverters in proportion to their rating during transients and in steady state. The following simulations are performed with devices of the same power to more easily compare the results.

3.5.2 Stability in Islanded Mode

In order to compare the response of the proposed algorithm with the conventional droop control, the following microgrid (Figure 3.18) is simulated, assuming islanded operation.

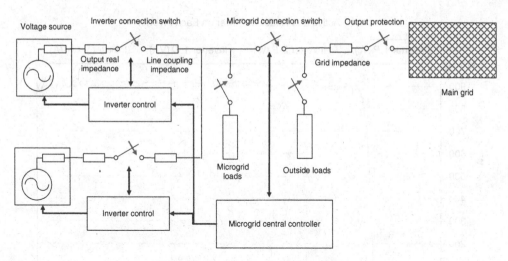

Figure 3.18 Simulated microgrid layout

The system comprises two inverters connected to a set of active and reactive loads. The inverters are coupled in parallel, and each inverter is composed of a three-phase voltage source, the three-phase connection switch, the actual output impedance and the line coupling impedance. Inverters are sized at 10 kW with a droop control characterized by 1 Hz difference between zero and maximum power output. The value of the output inductance in use is 0.15 Ω. Taking into account that the device nominal power is 10 kW, the output impedance, expressed in p.u. (per unit), is around 1% of the characteristic impedance. This rather low value is an adverse factor to the stability of the resulting system.

In this section and the next, the benefits of the proposed control are established at islanded operation, where the self synchronization and load sharing are important under different coupling conditions – resistive and inductive. Nevertheless, the control is also assessed when the system is connected to the grid, considering operations such as additional inverters being connected, synchronization and grid reconnection with subsequent disconnection, as well as the dynamics of the primary energy source.

3.5.2.1 Inductive Coupling

The first study case compares the performance of the droop control with and without the fictitious impedance. The value of the reactance considered in the simulations is 1–5% of the characteristic impedance of the devices.

In the droop-alone control scheme, simulations demonstrate that the system is more stable if the coupling inductance is high, around 5% in each inverter, 0.7 Ω and 0.8 Ω. The line should have a small resistive component to damp the high frequency (close to the fundamental) oscillations of power. The cut-off frequency of the low-pass filters may be reduced to eliminate the high-frequency oscillations but then the system may become unstable after exposing slower power oscillations in the region of the classic machine stability owing to the droop control. These latter oscillations are dependent on the droop constants and the value of the coupling reactance. For example, with smaller values of the inductance, system stability is

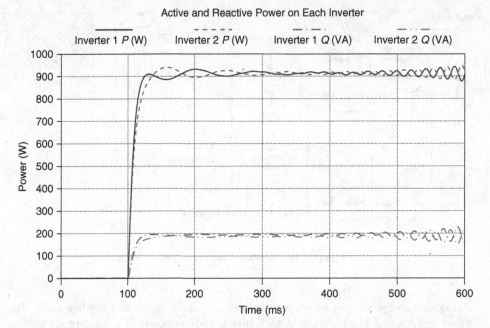

Figure 3.19 Output power of each inverter with totally inductive coupling and traditional drop control

difficult to achieve if the cut-off frequency of the filters is not increased. It becomes evident that combining measures to tackle these two stability issues is not an easy task in all practical situations. Using the fictitious impedance method, stability is easier to achieve in all situations.

In the following simulations, both inverters are synchronized at the nominal frequency (50 Hz) at time 0.1 seconds, with no initial power output and with a connected load of 1800 W, 400 VAr (inductive). Figure 3.19 shows the behavior of the control with purely inductive coupling. The control starts working properly, but a few milliseconds later the power starts to oscillate and the system becomes unstable. If a small resistive component, say $0.05\,\Omega$, is included in the coupling between the inverters, the oscillations die out. This is a small resistance that every LV coupling is expected to have, so it can be said that the system is stable.

The coupling reactances of the inverters are 0.7 and $0.8\,\Omega$ and the cut-off frequencies of the first-order filters are 20 Hz and 10 Hz for the current and power filter respectively.

When the coupling reactance is reduced in further simulations, it is shown that the system fails to reach stability. Therefore, the droop method alone does not seem robust enough to connect inverters in parallel under all conditions.

In addition, in the traditional droop method the need for high cut-off frequency of the filter causes the frequency of the system to vary very rapidly, reacting to load changes. In Figure 3.20, the output frequency of both inverters is plotted, showing rapid response and oscillations before reaching steady state.

With the fictitious impedance method the stability is easier to reach without the need to fine tune the control parameters. All simulations have been done with $1.4\,\Omega$ fictitious reactance. The cut-off frequency of the current low-pass filter is set at 5 Hz and the cut-off frequency of

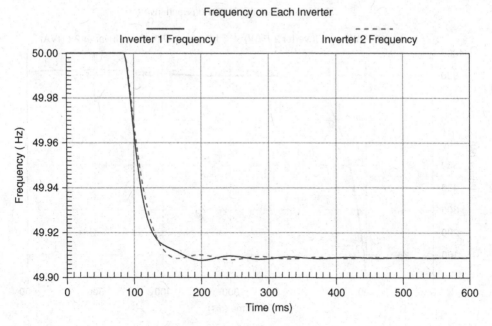

Figure 3.20 Output frequency of each inverter with inductive coupling and traditional droop control

the power low-pass filter is 1 Hz. Simulations have been done with different coupling reactances and the system proved stable in all cases. The system is stable even when the coupling between the inverters is purely reactive, and there is no resistance to damp the oscillations, provided that the fictitious impedance includes a resistive part. In Figure 3.21, it can be seen that the system response is slower than in the previous cases. Figure 3.22 illustrates the inverters frequency response; it is clear that the system frequency now moves gradually to the new balance point.

3.5.2.2 Resistive Coupling

The resistive coupling poses a big problem for the control of the parallel inverter units. In the following simulations each inverter is considered to be coupled with a resistance with value varying from 0 to 0.8 Ω, which means 0% to more than 5% of the characteristic impedance of the devices. When the traditional droop control is applied, stability is very difficult to reach and the system becomes unstable due to the cross-coupling of the control variables (Figure 3.23).

The use of the fictitious impedance helps to achieve stability in almost all of the cases. With no coupling impedance at all – that is, only with the device output impedance – the system is stable. When the coupling resistance is small the system remains stable, as shown in Figure 3.24, where the behavior of the system with coupling resistances of 0.1 Ω and 0.0 Ω, is shown.

If the coupling resistance is increased, to 0.3 Ω and 0.4 Ω, oscillations take more time to be damped, but the system is still stable. However, when the coupling resistance is increased to

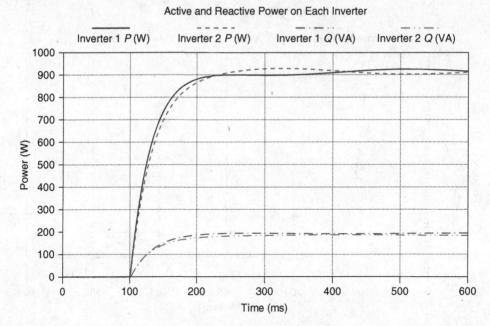

Figure 3.21 Output power of each inverter with inductive coupling and fictitious impedance method

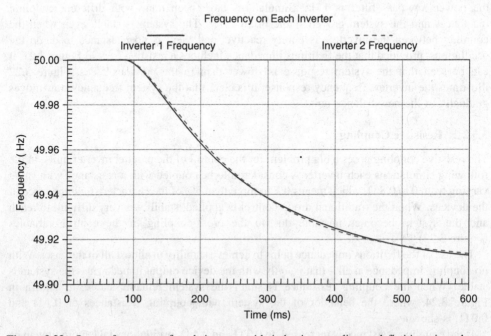

Figure 3.22 Output frequency of each inverter with inductive coupling and fictitious impedance method

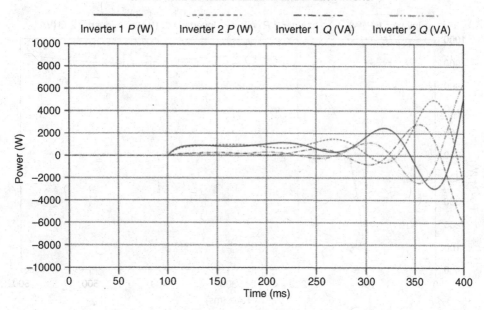

Figure 3.23 Output power of each inverter with resistive coupling and traditional drop control

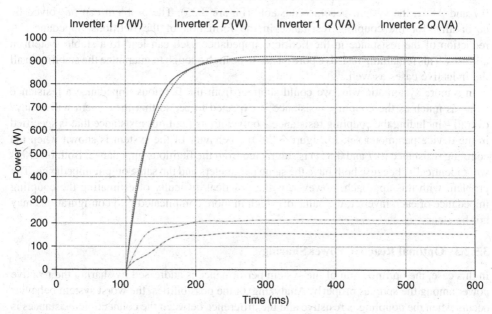

Figure 3.24 Output power of each inverter with resistive coupling and fictitious impedance method, with small resistance

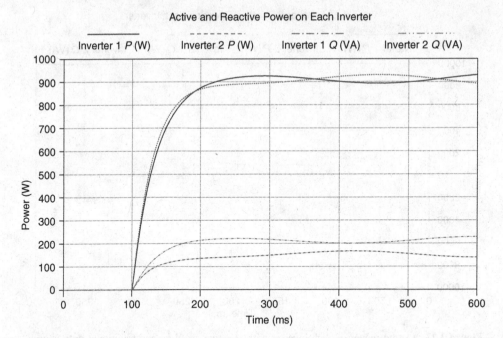

Figure 3.25 Output power of each inverter with resistive coupling and fictitious impedance method, with high resistance

0.7 and 0.8 Ω, the system becomes unstable (Figure 3.25). The problem can be solved by accounting for the coupling resistance in the formation of the fictitious impedance. A reduction of the resistance in the fictitious impedance used can lead to a stable system in all cases. This remaining resistive portion has to be high enough to guarantee the stability in all the inductive cases as well.

In a more systematic way, we could subtract from the fictitious impedance a resistance value as much as the coupling resistance is expected to be, so that the system will behave overall – including the coupling resistance – only with its fictitious resistance that is specified in the device parameter set. In Figure 3.26, the behavior of the system is shown when the coupling resistance (0.7 and 0.8 Ω) is subtracted from the fictitious impedance. Both inverters work identically, because both have the same parameters and the same output impedance. The problem with this approach, however, is the practical difficulty of estimating the coupling impedance of each inverter, the value of which in more complicated grid configurations may not be clear.

3.5.2.3 Optimal Reactive Power Sharing

In this case, the optimization of the system performance is addressed by sharing the reactive power among the sources properly. Analyzing all the possibilities, the worst system behavior occurs when the coupling is resistive and the difference between the connection resistances is significant. In these cases, the reactive power is generated mostly by the inverter with the smaller coupling resistance.

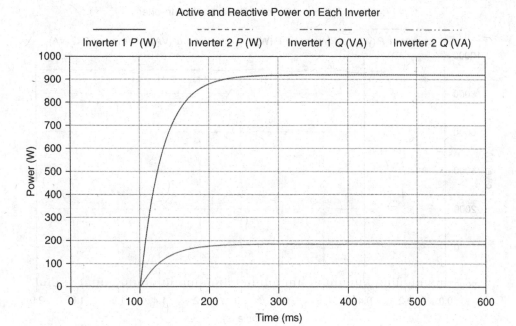

Figure 3.26 Output power of each inverter with resistive coupling and fictitious impedance method, with high resistance and compensated

The simulated microgrid consists of two similar inverters with a difference between the coupling resistances equal to $0.4\,\Omega$. Figure 3.27 displays the behavior of the system when the fictitious impedance is $1.4\,\Omega$. It can be seen that while the active power is equally shared, the reactive power is mainly generated by the inverter with the smaller coupling resistance. This behavior is only dependent on the connection impedance and the fictitious impedance. The difference between the two reactive powers increases when the fictitious impedance decreases. The reactive power productions are balanced when the voltage drop provoked by the active power across the extra resistance is equal to the voltage drop created by the difference of the reactive power across the fictitious impedance. This difference can be used by a central controller to estimate the coupling resistances of each inverter, or at least the difference between the inverter resistances. Figure 3.28 shows the performance of the system when the coupling resistance is compensated.

3.5.3 Stability in Interconnected Operation

In this section the progressive connection of different power sources, the synchronization process of the microgrid with the main grid and the subsequent reconnection and disconnection are studied from the point of view of system stability, proposing improvements in the control strategies.

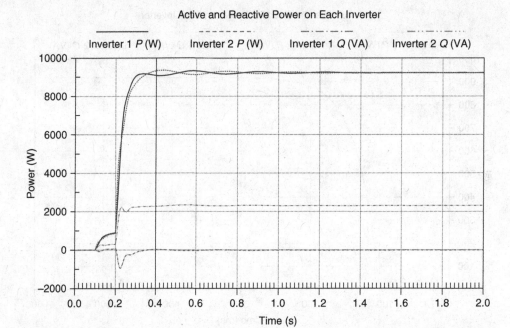

Figure 3.27 Output power of each inverter with unequal resistive coupling and $1.4\,\Omega$ fictitious impedance

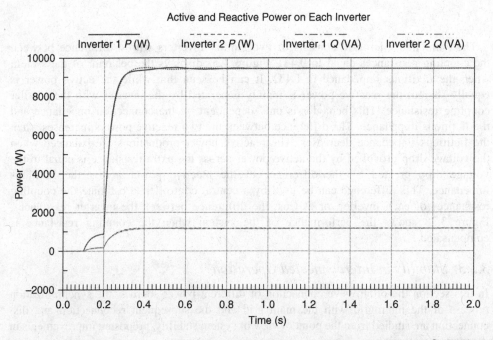

Figure 3.28 Output power of each inverter with unequal resistive coupling and $1.4\,\Omega$ fictitious impedance compensated

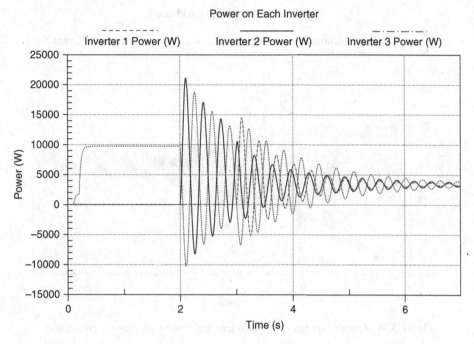

Figure 3.29 Output power of the inverters during their progressive connection

3.5.3.1 Progressive Connection of Microsources

At time $t = 0.1$ s, a first inverter is connected to a 2 kW load. The load initially has no voltage, so the inverter energizes the microgrid. At time $t = 0.2$ s, the load is increased to 10 kW. After 2 seconds, when the system has stabilized, a second inverter is connected to the network and the connection of a third inverter follows one second later. All inverters are similar, but a slight difference in their parameters is assumed, in order to study stability issues. It is assumed that each inverter, before connecting to the existing network, has synchronized its output voltage reference (magnitude and phase). The behavior of the system, as expressed by the output power of the three inverters is shown in Figure 3.29 and the inverter currents in Figure 3.30.

It can be seen that when a new inverter is connected to the microgrid, currents and powers flow from one inverter to the other, until the system reaches equilibrium. One reason for causing fluctuations is that when a new inverter is synchronized, its voltage angle is the same as the angle of the grid voltage, but the frequency is imposed by the power measurement (because of the P/f droop). If the two frequencies are not the same, the inverter frequencies fluctuate around the equilibrium point causing current oscillations. If the frequencies are equal at the first instant, the system moves smoothly to the steady state.

Two main alternatives can be considered to reduce oscillations:

- adapt the internal values of the inverter in order to achieve the same frequency of the inverter and the system during connection

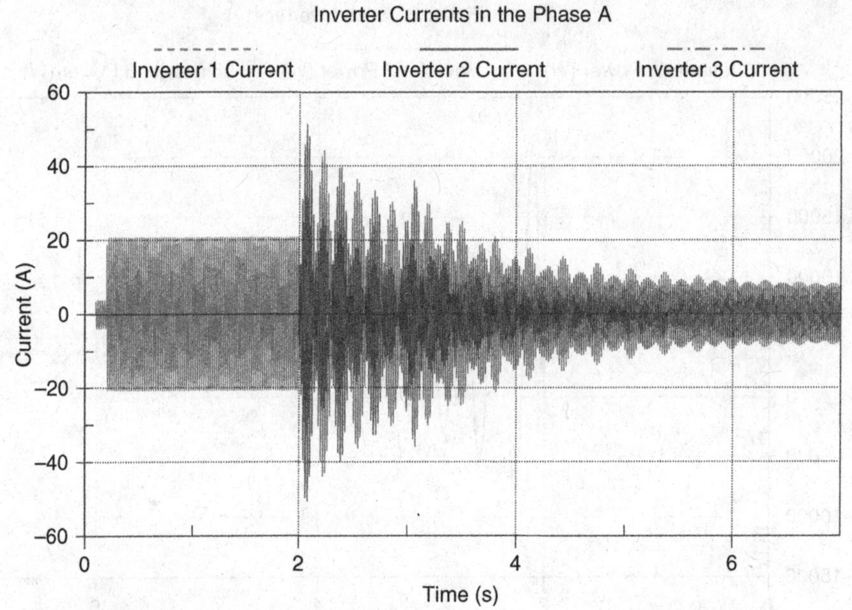

Inverter Currents in the Phase A

Figure 3.30 Output currents of the inverters during their progressive connection

- adjust the dynamic characteristics of the control system at the connection time to improve fluctuation damping

For example, it is possible to act on the P/f droop by adding an offset value to the droop characteristic to obtain the same frequency in the two inverters. Initially, and before connecting the new inverter, the grid frequency is calculated and compared with the inverter zero power frequency. The difference is added to the P/f droop curve when the new inverter connects, resulting in 0 kW power at its output. So the system is supplied by the already connected inverters, while the new one is synchronized with the island. The offset is gradually reduced to zero and the system reaches balance smoothly. Figure 3.31 shows the output powers and Figure 3.32 the currents.

The dynamic parameters of the inverters can be changed by temporarily replacing the coupling fictitious impedances and resistances, substituting the cut-off frequencies of the current and power filters, tuning the P/f droop slope or even modifying the digital filter transfer function. It is worth mentioning that the only parameter set that is to be adjusted is the one belonging to the inverter being connected.

The fictitious impedance can be employed to reduce the transients in power and current by reducing an initial high inductance value L_{Do}^* to the final lower value L_{Df}^* with a time constant T_{ST} according to $L_D^* = L_{Df}^* + \left(L_{Do}^* - L_{Df}^* \right) e^{-t/T_{ST}}$, as shown in Figure 3.33.

3.5.3.2 Transition Between Islanded and Connected Modes

Microgrid transitions from interconnected to islanded mode of operation and vice versa are likely to cause large mismatches between generation and loads, provoking severe frequency and voltage variations.

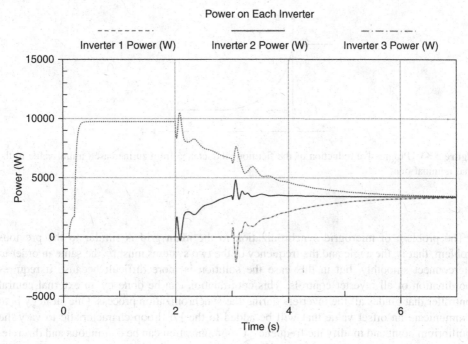

Figure 3.31 Output power of the inverters with offset technique during their progressive connection

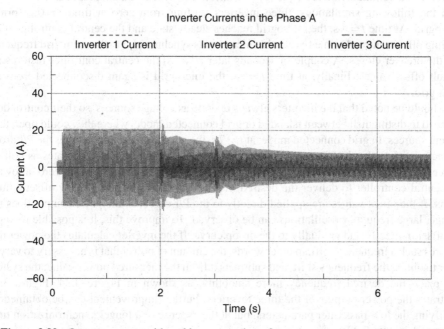

Figure 3.32 Output currents with sudden connection of new inverters using offset technique

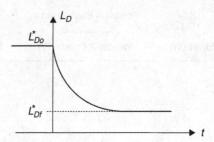

Figure 3.33 Progressive reduction of the fictitious inductance from an increased initial value to the final nominal one

The problem of microgrid synchronization to the main grid is similar to the previous problem, that is, the angle and the frequency of the two systems must be the same in order to interconnect smoothly, but in this case the solution is more difficult because it requires coordination of all inverter controls. This coordination can be done by an external central controller that guides all the inverters during the synchronization process. One strategy is to communicate an offset value that will be added to the P/f droop characteristic to vary the equilibrium point and modify the frequency. Communication can be continuous and discrete. With continuous communication (wired signal), the synchronization process can be faster and softer, while with discrete communication the offset value is renewed in successive periods of time. In the following simulations the offset value is communicated every half second, but in a real case the communication frequency will affect the synchronization speed.

In the following simulations, three inverters starting from zero at time $t = 0$ s, form a microgrid. At time $t = 1$ s, the microgrid reaches steady state and the central controller starts sending offset values. At time t $= 5$ s, the microgrid is synchronized to the main grid frequency and the breaker closes. A couple of seconds later ($t = 7$ s) the central controller restores the default offset values. Finally, at time $t = 9$ s, the microgrid is again disconnected from the main grid.

It should be noted that the inverters always operate as voltage sources, so their control does not need to distinguish between islanded or grid connected mode, where they could operate as current sources. In grid connected mode, at $t = 7$ s, the central controller changes the P/f droop characteristics to the default value. Taking into account that the inverter operates as voltage source and that network frequency is 50 Hz, the inverter power output goes to zero, allowing the central controller to deliver the desired import/export power through new offset adjustments. If the offset values are applied directly to the P/f curve, as the power output does not change, large frequency oscillations can be observed. To improve this, it is possible to apply the offset indirectly and gradually to the droop curve. If the inverter calculates the power that requires such a frequency shift, in other words, the amount of power that is necessary to vary in order to obtain the frequency shift, and subtracts it from the measured power value, the system will reach the desired frequency more smoothly, as shown in Figure 3.34. Figure 3.35 illustrates the power outputs of the three inverters. Further improvement can be obtained by modifying the low-pass filter parameters, but at the expense of a longer synchronization time (Figures 3.36 and 3.37).

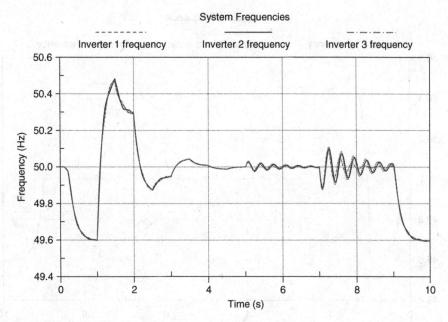

Figure 3.34 Inverter frequencies during transitions from islanded to interconnected mode of operation with indirect offset ($t < 5$ s, synchronization; $t = 5$ s, connection to the main grid; $t = 7$ s, removing of the synchronization offset; $t = 9$ s, disconnection from the main grid)

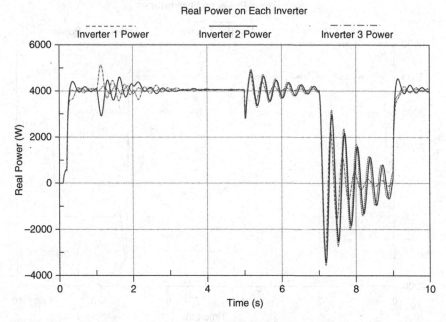

Figure 3.35 Inverter active power during transitions from islanded to interconnected mode of operation with indirect offset

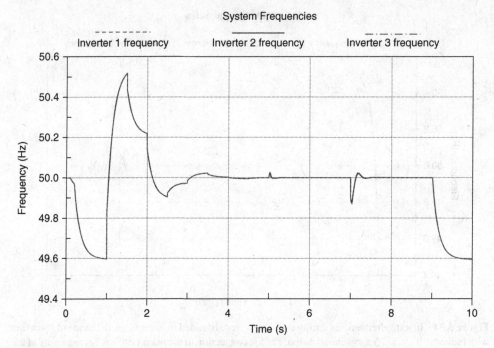

Figure 3.36 Inverter frequencies during transitions from islanded to interconnected mode of operation with indirect offset and modified low-pass filter function

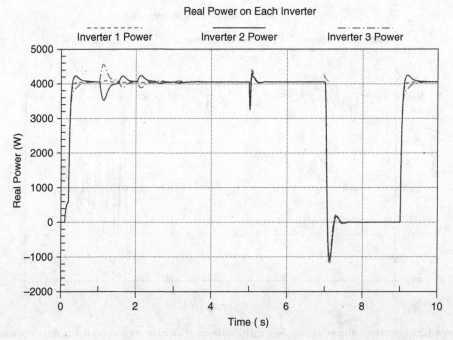

Figure 3.37 Inverter active power during transitions from islanded to interconnected mode of operation with indirect offset and modified low-pass filter function

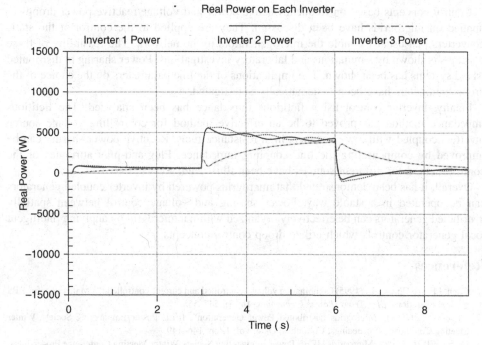

Figure 3.38   Inverter active powers of different sources during sudden load changes

Some power oscillations appear in the connection and disconnection process as the smoothing techniques used in the previous cases are only useful for the individual element reconnection. For a reconnection, in which no communication is used to notify to the inverters their new state, the power measurement low-pass filter can be modified to smooth the oscillations.

3.5.3.3 DER Characteristics

Clearly, not all microsources and storage devices have the same dynamic behavior, and their individual characteristics should be considered. The control algorithm must adapt to the dynamical behavior of the different sources, such as the inherent storage capability or their power change rate, taking advantage of the characteristics of each distributed source (Figure 3.38).

In general, the dynamic response of the inverters is related to their coupling impedances and their transfer functions. The inverter with smaller impedances will react first, while those with larger impedances will take a longer time. The oscillations can be damped by modifying the low-pass filter transfer functions.

3.6 Conclusions

This chapter has outlined the functions of local intelligent controllers of microsources in a microgrid. These controllers are responsible for voltage and frequency control in microgrids and are essential for stable microgrid operation.

Control concepts based on frequency/active power and voltage/reactive power droops as applied on microgrids have been discussed. They are applied to the control of the static converters that usually couple the microgenerators to the network. The feasibility of these concepts is shown by simulations and laboratory investigations. Power sharing in distributed island systems has been shown. The implications of the line parameters on the choice of the droop parameters have been systematically investigated.

Finally, inverter control using fictitious impedance has been analyzed. The fictitious impedance method has proved to be an effective method for controlling voltage source inverters coupled with low reactance, high resistance lines. Reactive power sharing can be improved by compensating the line coupling impedance. Plug-and-play attributes of the control method for different sources have been illustrated.

Overall, it has been demonstrated that microgrids powered by inverter coupled generators can be operated in a stable way. Power sharing and voltage control between spatially distributed generators can be effectively organized within a microgrid by applying intelligent local generator controls which utilize droop control concepts.

References

1. Chen, J.F. and Chu, C.-L. (1995) Combination voltage-controlled and current-controlled PWM inverters for UPS parallel operation. *IEEE Trans. Power Electronics*, **10** (5), 547–558.
2. Lasseter, B. (2001) "Microgrids Distributed Power Generation", IEEE Power Engineering Society Winter Meeting Conference Proceedings, Columbus, Ohio, vol. 1, pp. 146–149.
3. Lasseter, R.H. (2002) "Microgrids" IEEE Power Engineering Society Winter Meeting Conference Proceedings, New York, NY, vol. 1, pp. 305–308.
4. Engler, A. Vorrichtung zum gleichberechtigten Parallelbetrieb von ein- oder drei-phasigen Spannungsquellen, European patent: 1286444 B1 (2006) US patent: 6,693,809 B2 Feb 17, 2004 Japanese patent 2010-1102209.
5. Engler, A., Hardt, C., Strauß, P., and Vandenbergh, M. (October 2001) "Parallel Operation of Generators for Stand-Alone Single Phase Hybrid System", EPVSEC Conference, Munich.
6. Engler, A. (2002) *Regelung von Batteriestromrichtern in Modularen und Erweiterbaren Inselnetzen*, Publisher Verlag Dissertation.de, Berlin, ISBN: 3-89825-439-9.
7. Heuck, K. and Dettmann, K.D. *Elektrische Energieversorgung*, 3rd edn, Vieweg.
8. Burger, B. and Engler, A. (2000) Verfahren und Vorrichtung zur Bestimmung charakteristischer Grössen aus einem zeitlich periodischen Signal, German patent Nr. 199 49 997.7, 15.10.1999.
9. Engler, A. (2005) Applicability of droops in low voltage grids. *International Journal of Distributed Energy Resources*, **1** (1) 3–15.
10. Chiang, H.C., Yen, C.Y., and Chang, K.T. (2001) A frequency-dependent droop scheme for parallel control of Ups inverters. *J. Chin. Inst. Eng.*, **24** (6), 699–708.
11. Guerrero, J.M., Vasquez, J.C., Matas, J. *et al.* (2011) Hierarchical control of Droop – Controlled AC and DC microgrids – A general approach towards standardization. *IEEE Trans. Indus. Electronics*, **58** (1) 158–172.

4

Microgrid Protection

Alexander Oudalov, Thomas Degner, Frank van Overbeeke
and Jose Miguel Yarza

4.1 Introduction

One of the major technical challenges associated with a wide deployment of microgrids is the design of its protection system. Protection must respond both to the utility grid system and to microgrid faults. If the fault is on the utility grid, the desired response is to isolate the microgrid from the main utility as rapidly as necessary to protect the microgrid loads. The speed of isolation is dependent on the specific customer's loads on the microgrid, but it probably requires the development and installation of suitable electronic static switches. Electrically operated circuit breakers in combination with directional over-current protection is another possible option. If the fault is within the microgrid, the protection system isolates the smallest possible section of the distribution feeder to eliminate the fault [1]. A further segmentation of the microgrid during the isolated operation, that is, a creation of multiple islands or sub-microgrids, must be supported by microsource and load controllers.

Most conventional distribution protection is based on short-circuit current sensing. The presence of DERs might change the magnitude and direction of fault currents, and might lead to protection failures. Directly coupled rotating-machine-based microsources will increase short-circuit currents, while power electronic interfaced microsources can not normally provide any significant levels of short-circuit currents required. Some conventional over-current sensing devices will not even respond to this low level of over-current, and those that do respond will take many seconds to do so, rather than the fraction of a second that is required. Thus, in many operating conditions of microgrids, problems related to selectivity (false, unnecessary tripping), sensitivity (undetected faults) and speed (delayed tripping) of protection system may arise.

Microgrids: Architectures and Control, First Edition. Edited by Nikos Hatziargyriou.
© 2014 John Wiley & Sons, Ltd. Published 2014 by John Wiley & Sons, Ltd.
Companion Website: www.wiley.com/go/hatziargyriou_microgrids

Issues related to protection of microgrids have been addressed in several publications [2–5]. The major problems can be summarized as:

- changes in the value and direction of short-circuit currents, depending on whether a distributed generator is connected or not,
- reduction of fault detection sensitivity and speed in tapped DER connections,
- unnecessary tripping of utility breaker for faults in adjacent lines, due to fault contribution of DER,
- increased fault levels may exceed the capacity of the existing switchgear,
- auto-reclosing and fuse-saving of the utility line breaker policies may fail,
- reduced fault contribution of inverter based DER on protection system performance, especially when isolated from the utility grid,
- conflict between the feeder protection and utility requirements for fault ride through (FRT) which is included in the grid codes of many countries with a large penetration of DER,
- effect of closed-loop and meshed distribution network topologies with DER.

This chapter presents a protection solution for microgrids that can overcome some of the above problems. This is based on automatic adaptive protection, that is, change of protection settings depending on the microgrid configuration based on pre-calculated or on real-time calculated settings. The increase in the fault current level helped by a dedicated device, especially in isolated operation of microgrids dominated by power electronics interfaced DER, is also analyzed. Finally, the possible use of fault current limitation is discussed.

4.2 Challenges for Microgrid Protection

4.2.1 Distribution System Protection

Generally a distribution system (including a microgrid) is divided into local protective zones which are covered either by a network (overhead lines and cables) or apparatus (buses, transformers, generators, loads etc.) protection (Figure 4.1).

Requirements which provide a basis for design criteria of a distribution protection system are known as "3S" which stands for:

- sensitivity – protection system should be able to identify an abnormal condition that exceeds a nominal threshold value.
- selectivity – protection system should disconnect only the faulted part (or the smallest possible part containing the fault) of the system in order to minimize fault consequences.
- speed – protective relays should respond to abnormal conditions in the least possible time in order to avoid damage to equipment and maintain stability.

"3S" can be extended by:

- dependability – protection system must operate correctly when required to operate (detect and disconnect all faults within the protected zone) and shall be designed to perform its intended function while itself experiencing a credible failure.

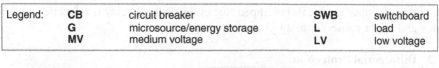

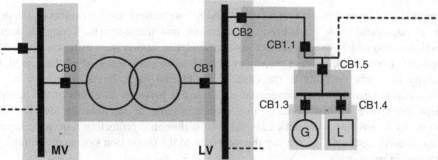

Figure 4.1 Protection zones of different MV and LV circuit breakers with over-current relays for grid elements and apparatuses

- security – protection system must not operate when not required to operate (reject all power system events and transients that are not faults) and shall be designed to avoid misoperation while itself experiencing a credible failure.
- redundancy – protection system has to care for redundant function of relays in order to improve reliability. Redundant functionalities are planned and referred to as backup protection. Moreover, redundancy is reached by combining different protection principles, for example distance and differential protection for transmission lines.
- cost – maximum protection at the lowest cost possible.

4.2.1.1 Over-Current and Directional Over-Current Protection

The protection of distribution grid, where feeders are radial with loads tapped-off along feeder sections, is usually designed assuming a unidirectional power flow. It is based on a detection of high fault currents using fuses, thermo-magnetic switches and molded case circuit breakers with standard over-current (OC) relays (ANSI 51) with time-current discriminating capabilities. More sophisticated directional OC relays (ANSI 67) are used for the protection of ring and meshed grids.

4.2.1.2 Distance Protection

Some utilities install distance relays (ANSI 21) for line protection. The distance relay compares the fault current against the voltage at the relay location to calculate the impedance from the relay to the faulted point. As a rule of thumb, a distance relay has three protection zones: zone 1 covers 80–85% length of the protected line, zone 2 covers 100% length of the protected line plus 50% of the next line, zone 3 covers 100% length of the protected line plus 100% of the second line, plus 25% of the third line. If a fault occurs in the operating zone of the distance relay, the measured impedance is less than the setting, and the distance relay operates to trip the circuit breaker. Unfortunately, the distance protection can be affected by

DERs and loads, since the measured impedance of the distance relay is a function of in-feed currents and might cause the relay to operate incorrectly.

4.2.1.3 Differential Protection

Differential over-current protection relays (ANSI 87) are mainly used to protect an important piece of equipment such as distributed generators and transformers. Today, differential protection is also widely used to protect underground distribution lines using communication (pilot wires, fiber optics, radio or microwave etc.) between line terminals. It has the highest selectivity and only operates in the case of an internal fault, but it requires a reliable communication channel for instantaneous data transfer between terminals of the protected element (pilot wire, optical fibres or free space via radio or microwave). Because of its vulnerability to possible communication failures, differential protection requires a separate backup protection scheme, increasing the total cost of the protection system and limiting its application in microgrids.

Although several protection principles can be used in LV distribution grids, over-current protection is the most dominant. Therefore, the following sections focus on over-current protection in microgrids.

4.2.2 Over-Current Distribution Feeder Protection

Over-current protection detects the fault from the high value of the current flowing to the fault. In modern digital (microprocessor based) relays a tripping short-circuit current I_k can be set in a wide range (e.g. 0.6–15 times the rated current of a circuit breaker I_n). If a measured line current is above the tripping setting, the relay operates to trip the CB on the line, with a delay defined by a coordination study and compatible with a locking strategy used (no locking, fixed hierarchical locking, directional hierarchical locking).

A typical shape of over-current trip curve of a modern circuit breaker is shown in Figure 4.2. The trip curve consists of an inverse time part L (protection against overloads), a constant time

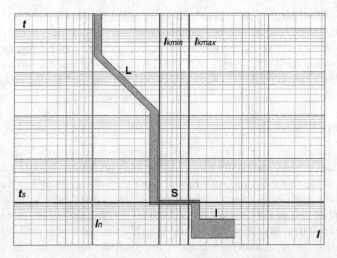

Figure 4.2 Typical time-current tripping curve of an LV circuit breaker with electronic release

delay part S (protection against short circuit with short-time delay trip) and an instantaneous part I (instantaneous protection against short circuit). The S part may consist of several steps. It is common today that in modern digital over-current relays, all parts of the trip curve are adjustable in a wide range by small increment steps.

4.2.3 Over-Current Distribution Feeder Protection and DERs

Most of the microgenerators and energy storage devices installed at LV distribution grids such as solar (photovoltaic) panels, micro-wind and micro-gas turbines (usually combined heat and power), are not capable for supplying power directly to the grid and have to be interfaced by means of power electronics (PE) components. The use of PE interfaces leads to a number of challenges in microgrid protection, especially in islanded mode.

Figure 4.3 shows a typical microgrid comprising two feeders connected to the LV bus and to the MV bus via a distribution transformer. Each feeder has three switchboards (SWB). Each SWB has star or delta configuration and connects DERs and loads to the LV feeder. In the following, we analyze two external (F1, F2) and two internal (F3, F4) microgrid faults. All LV circuit breakers (from CB1 to CB6.5) may have different ratings but are equipped with a conventional over-current protection and are used for segmenting the microgrid.

In general, protection issues in a microgrid can be divided in two groups, depending on the microgrid operating state (Table 4.1). The table also shows the importance of "3S" (sensitivity, selectivity and speed) requirements for different cases, which provides a basis for design criteria of a microgrid protection system.

The next sections provide more details on each case of Table 4.1.

4.2.4 Grid Connected Mode with External Faults (F1, F2)

With fault F1, a main grid (MV) protection clears the fault. If sensitive loads are connected to the microgrid, it could have to be isolated by CB1 as fast as 70 ms (depending on the voltage sag level in the microgrid) [6]. Also, the microgrid has to be isolated from the main grid by

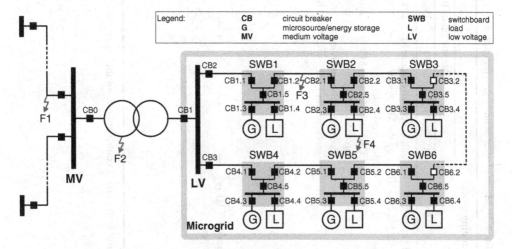

Figure 4.3 External and internal fault scenarios in a microgrid

Table 4.1 Major classes of microgrid protection problems

Operating mode	Fault location			
	External faults (main grid)		Internal faults (microgrid)	
	MV feeder, busbar (F1)	Distribution transformer (F2)	LV feeder (F3)	LV consumer (F4)
Grid connected (CB1 is closed)	Fault is normally managed by the MV system. Microgrid isolation by CB1 in case of no MV protection tripping. Possible fault sensitivity problems for CB1.[a]	Fault is normally managed by MV system (CB0). CB1 is opened by follow-me function of CB0. In case communication fails, possible fault sensitivity problem for CB1.[a]	Disconnect the smallest portion of microgrid (CB1.2 and CB2.1). CB1.2 is opened by fault current from the grid (high level). Low level of a reversed fault current from feeder's end may cause sensitivity problems for CB2.1.[a] In this case, a follow-me function of CB1.2 can open CB2.1. In case communication fails, possible fault sensitivity problems for CB2.1.[a]	Faulted load is isolated by CB2.4 or a fuse. In case of no tripping, the SWB is isolated by CB2.5 and the local DER is cut off. Sensitivity or selectivity problems unlikely.
Islanded (CB1 is open)	—	—	Disconnect the smallest portion of microgrid (CB1.2 and CB2.1). Low level of fault currents from both directions may cause sensitivity problems for CB1.2 and CB2.1.[a]	Faulted load is isolated by CB2.4 or a fuse. In case of no tripping, the SWB is isolated by CB2.5 and the local DER is cut off. Sensitivity or selectivity problems unlikely.

[a] Low fault current contribution from the microgrid in case of PE interfaced DERs.

CB1, in case of MV protection tripping failure. A detection of F1 with a generic OC relay can be problematic, in case most of the DERs in the microgrid are connected via PE interfaces with built-in fault current limitation (i.e. there is no significant rise in current passing through CB1). Typically, these units are capable of supplying 1.1 to $1.2*I_{DERrated}$ (rated current of microgenerator) to a fault, unless the converters are specifically designed to provide high fault currents. This is much lower than a short-circuit current supplied by the main grid. A directional OC relay in CB1 is the only feasible solution, if current is used for fault detection. In order to increase relay sensitivity, the pick-up current setting is defined as the sum of fault current contributions from the connected DER units (4.1). DERs that have to be taken into consideration are the subset of units that contribute to the short-circuit current on the defined direction.

$$I_{k\,min} = \sum_1^n k_{DER} * I_{rDER} \qquad (4.1)$$

I_{rDER} is the rated output current of a particular DER and k_{DER} is its fault current contribution coefficient, set at 1.1 for PE interfaced DER and at 5.0 for synchronous DER units [7]. This value will vary for a large number of different types of DER. Thus, the setting has to be continuously monitored and adapted when microgrid generation undergoes changes (number and type of connected DER).

Alternatively, voltage sag (magnitude and duration) or/and system frequency (instantaneous value and rate of change) can be used as indicators for tripping CB1. It may be also required that the energy sources of the microgrid stay connected and supply reactive power to support the utility grid during the fault conditions. Depending on the particular grid code the duration of support depends on the level of residual voltage and may vary from a hundred milliseconds to several seconds.

With fault F2, the distribution transformer OC protection clears the fault by opening CB0. CB1 is opened simultaneously by a "follow-me" function (hardware lock) of CB0. On hardware lock failure, a possible fault sensitivity problem can arise as in the case of fault F1. Typical solutions are similar to the F1 case: directional adaptive OC protection, under-voltage and under-frequency protection and various islanding detection methods [8].

4.2.5 Grid Connected Mode with Fault in the Microgrid (F3)

With fault F3, microgrid protection must disconnect the smallest possible portion of the LV feeder by CB1.2 and CB2.1. CB1.2 is opened due to a high short-circuit current supplied by the main MV grid. If CB1.2 fails to trip, fault F3 must be cleared by CB1.1 which is a backup protection for CB1.2. However, the sensitivity of the OC protection relay in CB1.1 can be potentially disturbed, if a large synchronous DER (e.g. diesel generator) is installed and switched on at SWB1 (i.e. between CB1.1 and the fault F3). In this case, the fault current passing through CB1.1 with DER will be smaller than without DER, while the fault current at F3 will be higher due to additional DER contribution. This effect is known as protection blinding (the larger the synchronous DER the greater the effect – in some cases the fault current seen by CB1.1 can be reduced by more than 30%) and may result in a delayed CB1.1 tripping, because of the fault current transition from a definite-time part to an inverse-time part of the relay tripping characteristic (Figure 4.2). A delayed fault tripping will lead to an unnecessary disconnection of DER at SWB1 (usually low-power diesel generators have very low inertia and will run out of step in case of a delayed fault tripping). This issue can be solved by proper coordination of microgrid and DER protection systems. Another option is to adapt the protection settings with regard to current operating conditions (DER status).

If CB1.2 operates faster than CB2.1, which is very likely, it will island a part of the microgrid which will be connected to the fault F3. If it is possible to balance generation and load in the islanded segment of the microgrid (microsources are capable of supplying loads directly or after load shedding), it is expedient to isolate that group of microsources and loads from the fault F3 by opening CB2.1 and possibly closing CB3.2–6.2. However, a reversed and low-level short-circuit current in case of PE interfaced DER will cause a sensitivity problem for CB2.1 similar to the one described in the case of fault F1. Possible solutions include directional adaptive OC protection and a follow-me function of CB1.2, which opens CB2.1 (in this case, a communication failure may cause sensitivity problems for CB2.1).

4.2.6 Grid Connected Mode with Fault at the End-Consumer Site (F4)

With fault F4, a high short-circuit current is supplied to the fault from the main grid, together with a contribution from DER, which will lead to tripping of CB2.4. Frequently, there is a fuse instead of a CB, which is rated in such a way that a shortest possible fault isolation time is guaranteed. In case of no tripping, the SWB2 is isolated by CB2.5 and local DER is cut off. No sensitivity or selectivity problems are foreseen in this scenario.

4.2.7 Islanded Mode with Fault in the Microgrid (F3)

The microgrid might operate in islanded mode when it is intentionally disconnected from the main MV grid by CB1 (full microgrid) or a CB along the LV feeder (a segment of the microgrid). This operating mode is characterized by absence of the high short-circuit current supplied by the main grid. Generic OC relays should be replaced by directional OC relays, because fault currents flow from both directions to the fault F3. If CB1.2 and CB2.1 use setting groups chosen for the grid-connected mode, they will have a sensitivity problem with detecting and selectively isolating the fault F3 and tripping within acceptable time frame for PE interfaced DER (the fault current could shift from a definite-time part to an inverse-time part of the relay tripping characteristic). There are two possible ways to address this problem:

- Install a source of high short-circuit current (e.g. a flywheel or a super-capacitor) to trip CBs or blow fuses with settings or ratings for the grid-connected mode. However, a short-circuit handling capability of PE interfaces can be increased only by increasing the respective power rating or by having extensive cooling, which both lead to higher investment cost [9]. This solution is discussed in Section 4.4.
- Install an adaptive microgrid protection using online data about the microgrid topology and status of available microsources or loads. This solution is discussed in Section 4.3.

4.2.8 Islanded Mode and Fault at the End-Consumer Site (F4)

With fault F4, a low short-circuit current is supplied to the fault from the local DERs. There is no grid contribution to the fault current level. However, CB2.4 settings selected for the main grid-connected mode are just slightly higher than the rated load current. This ensures that the end-customer site will be disconnected, even if only PE-interfaced DERs are operating in the microgrid. If there is no tripping, the SWB2 must be isolated by CB2.5 using a directional OC relay. Similar to the grid-connected mode, there are no sensitivity or selectivity problems foreseen in the islanded mode for a fault at the end-consumer site.

4.3 Adaptive Protection for Microgrids

4.3.1 Introduction

As discussed in Section 4.2, time over-current protection is the most common method employed in MV and LV distribution grids, as it is quite simple to configure and is an economically attractive option. In networks with interconnected DER, short-circuit currents detected by over-current protection relays depend on the point of connection and the type and size (feed-in power) of the DER units. In these networks directions and magnitudes of short circuit currents will vary almost continuously, since the operating conditions in a microgrid are constantly changing due to the intermittent DERs (wind and solar) and periodic load variation. Also, the network topology can be regularly changed in order to achieve loss minimization or other economic or operational targets. In addition, controllable islands of different size can be formed, as a result of faults in the utility grid or inside a microgrid. In such circumstances coordination between relays may fail and generic OC protection with a single setting group in each relay may become inadequate, so it will not guarantee a selective operation for all possible faults. Therefore, it is essential to ensure that protection settings chosen for OC protection relays dynamically follow changes in grid topology and location, type and amount of distributed energy resources. Otherwise, unwanted or uncoordinated operation – or failure to operate when required – may occur. To solve the problems originating from the interconnection of DERs the protection relay settings can be adapted in a flexible way combined with the utilization of over-current protection relays with identification of current direction. Adaptation of settings could mean a continuous adaptation of relay characteristic settings or a switching of relay settings groups.

Adaptive protection is defined as "an online activity that modifies the preferred protective response to a change in system conditions or requirements in a timely manner by means of externally generated signals or control action" [10]. Technical requirements and suggestions for a practical implementation of an adaptive microgrid protection system, as seen by the authors, are:

- use of digital directional OC relays, since fuses or electromechanical and standard solid state relays (especially for selectivity holding) do not provide the flexibility for changing the settings of tripping characteristics, and they have no current direction sensitivity feature,
- digital directional over-current relays, which must provide the possibility for using different tripping characteristics (several settings groups, e.g. modern digital OC relays for LV applications have 2–4 settings groups) which can be parameterized locally or remotely, automatically or manually,
- use of new/existing communication infrastructure (e.g. twisted pair, power line carrier or radio) and standard communication protocols (Modbus, IEC61850, etc.), such that individual relays can communicate and exchange information with a central computer or between different individual relays quickly and reliably to guarantee a required application performance.

The communication time lag and the maximum time lag are not critical values for adaptive protection because the communication infrastructure is used to collect information about the microgrid configuration and to change accordingly the relay settings only prior to the fault. The interlock, if required, is done by means of physical point-to-point connection. On master–slave protocols like Modbus, the changes of configuration have to be identified with a maximum delay of 1–10 seconds depending on the dimensions of the network, and the protection system reconfiguration has to be completed in times of the same order, provided that basic backup protection functions are present during the transitory phase. On peer-to-peer protocols like IEC61850, the changes in the network configuration trigger protection reconfiguration. The accepted delays are equivalent to the previous ones.

The following sections present design of **dynamic adaptive** microgrid protection based on pre-calculated and "trusted" setting groups. An alternative design based on real-time calculated settings is also discussed. As with the microgrid energy management applications discussed in Chapter 2, the implementation of an adaptive protection system can be deployed with a centralized or decentralized approach, depending on the distribution of functionalities. Each approach requires a different communication architecture. The following sections focus on the centralized approach.

4.3.2 Adaptive Protection Based on Pre-Calculated Settings

The principles of a centralized adaptive protection system can be explained by the following example. Let's consider the microgrid of Figure 4.3, which includes a microgrid central controller (MGCC) and a communication system, in addition to primary switching equipment, as shown in Figure 4.4. The function of the MGCC can be carried out by a programmable logic controller (PLC), a station computer or a generic PC, located in a distribution secondary substation (MV/LV). Electronics make each CB with an integrated directional OC relay capable of exchanging information with the MGCC master–slave scheme, where the MGCC is a master and the CBs are slaves, via the serial communication bus RS-485 and using the standard industrial communication protocol Modbus [11]. By polling individual relays, the

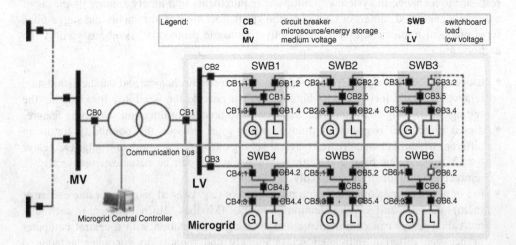

Figure 4.4 A centralized adaptive protection system

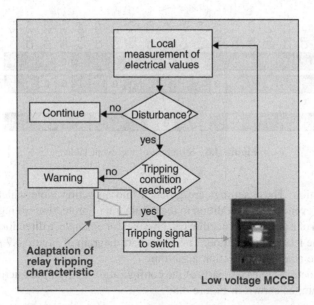

Figure 4.5 Local over-current protection function inside circuit breaker

MGCC can read data (electrical values, status) from CBs and, if necessary, modify relay settings (tripping characteristics).

When a fault happens, each individual relay takes a tripping decision locally (independently of the MGCC) and performs in accordance with the algorithm shown in Figure 4.5. The main goal of the adaptation module is to maintain settings of each OC relay regarding the current state of the microgrid (both grid configuration and status of DERs are taken into consideration).

The adaptation module in the MGCC is responsible for a periodic check and update of relay settings and consists of two main components:

- pre-calculated information during offline fault analysis of a given microgrid,
- an online operating block.

A set of meaningful microgrid configurations as well as feeding-in states of DERs (on/off) is created for offline fault analysis and is called an event table. Each record in the event table has a number of elements equal to the number of monitored CBs in the microgrid (some elements may have higher priority than others, such as the central CB which connects LV and MV grids) and is binary encoded, that is, element = 1 if a corresponding CB is closed and 0 if it is open (Figure 4.6).

Next, fault currents passing through all monitored CBs are calculated by simulating different short-circuit faults (three-phase, phase-to-ground etc.) at different locations of the protected microgrid at a time. During repetitive short-circuit calculations, the topology or the status of a single DER or load is modified between iterations. As different fault locations for different microgrid states are processed, the results (the magnitude and direction of fault current seen by each relay) are saved in a specific data structure.

Based on these results, suitable settings for each directional OC relay and for each particular system state are calculated in a way that guarantees a selective operation of

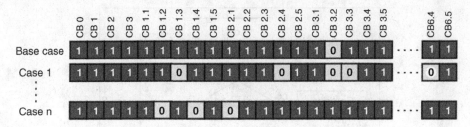

Figure 4.6 Structure of the event table

microgrid protection. These settings are grouped into an action table which has the same dimension as the event table. In addition to the regulation of protection settings, other actions such as activation of a protection function can be done, for example, a directional interlock can be activated in the islanding situation. The flow chart diagram of Figure 4.7 summarizes the offline part of the adaptive protection algorithm.

The event and action tables are parts of the configuration level of the microgrid protection and control system, as shown in Figure 4.8:

- *External field level* represents energy market prices, weather forecasts, heuristic strategy directives and other utility information.
- *Management level* includes historic measurements and the distribution management system (DMS).
- *Configuration level* consists of a station computer or PLC situated centrally (substation) or locally (switchboard) which is able to detect a system state change and send a required action to hardware level.

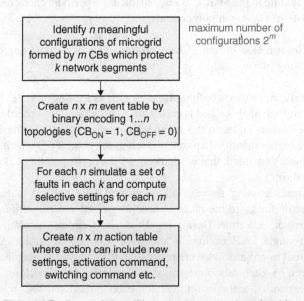

Figure 4.7 Steps of the offline adaptive protection algorithm

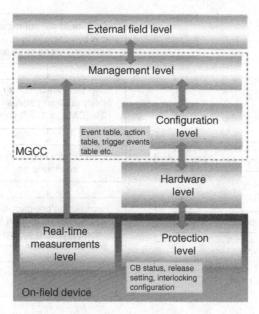

Figure 4.8 Microgrid protection and control architecture

- *Hardware level* transmits a required action from the configuration level to in-field devices by means of a communication network. In the case of a large microgrid, this function can be divided between several local controllers that communicate only selected information to the central unit.
- *Protection level* may include information, such as CB status, release settings and interlocking configuration. Together with the *real-time measurements level* they are residing inside in-field devices.

During online operation, the MGCC monitors the microgrid state by polling individual directional OC relays. This process runs periodically or is triggered by an event (tripping of CB, protection alarm etc.), as shown in Figure 4.9. The microgrid state information received by the MGCC is used to construct a status record that has a similar dimension to a single record in the event table. The status record is used to identify a corresponding entry in the event table. Finally, the algorithm retrieves the pre-calculated relay setting groups from the corresponding record in the action table and uploads the settings to in-field devices via the communication system. Alternatively, it may send commands to protection devices in the field, to switch between setting groups pre-loaded in the memory of these devices during system configuration.

The following examples illustrate the operation of the centralized adaptive protection system described in the previous paragraphs.

4.3.3 Microgrid with DER Switched off, in Grid-Connected Mode

The microgrid topology of Figure 4.10 is used as a base case and provides the first entry in the event table. The following Table summarizes its electrical characteristics.

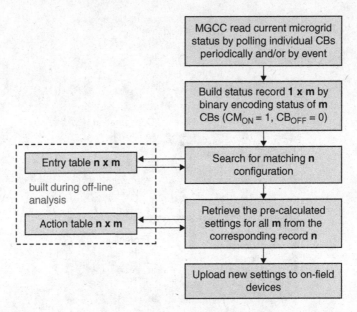

Figure 4.9 Steps of online adaptive protection algorithm with available look-up tables (event and action tables)

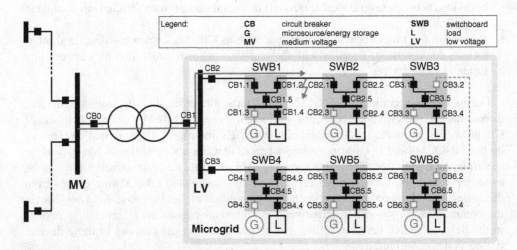

Figure 4.10 Scenario A1: the microgrid with DERs switched off, connected to the MV distribution grid

Parameters of utility grid	Value
Rated voltage, V	6000
Short-circuit power, MVA	500

Parameters of distribution transformer	Value
Rated voltage primary/secondary, V	6000/400
Rated power, kVA	630
Vcc, %	4
LV distribution system	TN-S

Parameters of cables	Value
TYPE	EPR/XLPE
Cross-section phase, mm^2	3×185
Cross-section neutral, mm^2	95
Nominal current, A	750
Resistance at 20 °C phase/neutral, mΩ	8.34/16.24
Inductance at 20 °C phase/neutral, mΩ	6.17/6.25
Length, meters	150

Parameters of feeder circuit breakers	Value
Rated voltage, V	400
Rated current, A	800

Parameters of loads	Value
Rated voltage, V	400
Rated current, A	100
Rated power factor, cosφ	0.9

Parameters of synchronous DERs	Value
Rated voltage, V	400
Rated apparent power, kVA	160
Rated power factor, cosφ	0.8
Direct-axis sub-transient reactance Xd″, %	9.6
Quadrature-axis sub-transient reactance Xq″, %	10.2
Direct-axis transient reactance Xd′, %	21
Direct-axis synchronous reactance Xd, %	260
Negative sequence reactance X2, %	9.8
Zero sequence reactance X0, %	2.1
Direct-axis sub-transient short circuit time constant Td″, ms	11
Direct-axis transient short circuit time constant Td′, ms	85

We assume that each protection device has a similar shape to the over-current protection trip curve (Figure 4.2). In order to provide a selective operation of different circuit breakers, we used different time delays t_s of the constant time delay part S in the range between I_{kmin} (the expected minimum short-circuit current) and I_{kmax} (the expected maximum short-circuit current). CB1 is the closest circuit breaker to the source and has the longest time delay t_s. The

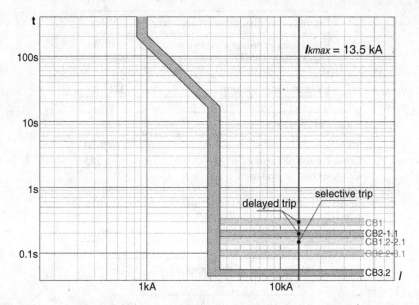

Figure 4.11 Trip curves for circuit breakers in the upper feeder (identical for CBs in the lower feeder) in the base case and a tripping sequence in scenario A1

most distant, CB3.2 and CB6.2, have the shortest time delay t_s (Figure 4.11). The instantaneous tripping part, I, is removed from all curves as a simplification.

The microgrid topology and suitable OC protection settings t_s for all CBs (calculated during the offline fault analysis [12]) in the base case are shown in Table 4.2, based on Figures 4.10 and 4.11. DER and load protection settings are not set here, but the information on DER and load status (on/off) is required for correct operation of the microgrid adaptive protection.

In case of a fault in the cable between SWB1 and SWB2 (Figure 4.10) all CBs between the fault and the LV busbar see the fault current supplied by the main MV grid (box in Table 4.2), but only CB1.2 will trip after $t_s = 150$ ms (Figure 4.11), and CB2.1 will be opened by the follow-me function of CB1.2, in order to avoid connecting the fault to the healthy feeder by closing CB3.2 and CB6.2. Other CBs that see the fault are delayed (in absence of a logic discrimination auxiliary connection). However, this method is limited by a number of discriminating time steps (a maximum t_s is recommended to be less than 800 ms) and is only suitable for feeders with a small number of switchboards.

If we assume that SWB2 and SWB3 are resupplied via SWB6 (CB3.2 and CB6.2 are closed) after the fault between SWB1 and SWB2 is selectively eliminated (CB1.2 and CB2.1 are open) a selectivity problem may appear if we use the base case protection settings of Table 4.2.

For example, if the second fault appears between SWB2 and SWB3 (Figure 4.12), it will be eliminated by CB3.2 and CB6.2 ($t_s = 50$ ms) instead of CB3.1 ($t_s = 100$ ms) and the load in SWB3 will be unnecessarily tripped as shown in Figure 4.13.

Selectivity can be improved by:

- application of directional over-current relays,
- modification of protection settings of non-directional OC relays.

Table 4.2 Scenario A1 – status of circuit breakers: 1 = close, 0 = open and over-current protection settings t_s in seconds. Box and numbers show CBs that see the fault in Figure 4.10.

Upper feeder	CB1	CB2	CB1.1	CB1.2	CB2.1	CB2.2	CB3.1	CB3.2
	1	1	1	1	1	1	1	0
	0.3	0.2	0.2	0.15	0.15	0.1	0.1	0.05

Lower feeder	CB3	CB4.1	CB4.2	CB5.1	CB5.2	CB6.1	CB6.2
	1	1	1	1	1	1	0
	0.2	0.2	0.15	0.15	0.1	0.1	0.05

DER + load	CB1.3	CB1.4	CB1.5	CB2.3	CB2.4	CB2.5	CB3.3	CB3.4	CB3.5
	0	1	1	0	1	1	0	1	1

DER + load	CB4.3	CB4.4	CB4.5	CB5.3	CB5.4	CB5.5	CB6.3	CB6.4	CB6.5
	0	1	1	0	1	1	0	1	1

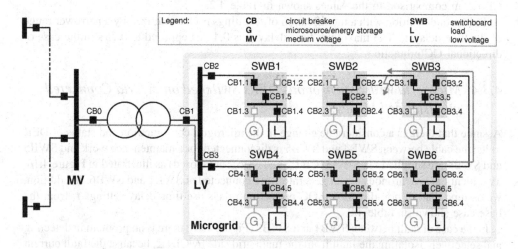

Figure 4.12 Scenario A2: the microgrid with all DERs switched off, connected to the MV distribution grid

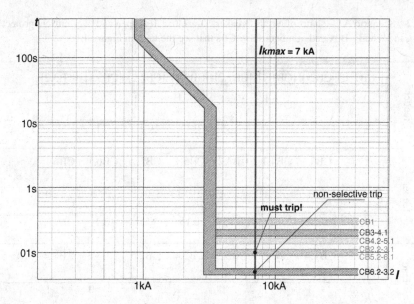

Figure 4.13 Base case trip curves and a tripping sequence in scenario A2 with non-directional OC protection

In the case of directional over-current protection, each relay will have two t_s settings, one in each direction (clockwise and counter-clockwise, or forward and backward), as shown in Table 4.3. In this case a selective protection operation is guaranteed (Figure 4.14) and SWB3 will remain connected after the fault is eliminated.

In the second case, we modify the S part of the base case trip curves by adjusting t_s settings (Figure 4.15). Second records in the event and action tables are created during the offline fault analysis (Table 4.4).

In particular, we can observe lower t_s for CB2.2 and CB3.1 and higher t_s for CB3.2 and CB6.2, in comparison to the values shown in Table 4.2.

The second solution with a modification of t_s settings is characterized by a narrower range of the time delay t_s. The maximum time delay t_s is 0.4 s as opposed to 0.75 s in the case of directional OC protection.

4.3.4 Microgrid with Synchronous DERs Switched on in Grid Connected and Islanded Modes

Assume that there is a considerable change in the microgrid configuration and status of DER units: the cable between SWB4 and SWB5 is disconnected for a maintenance work and SWB5 and SWB6 are supplied via SWB3 (CB3.2 and CB6.2 are closed) as illustrated in Figure 4.16. Two identical synchronous diesel generators are connected at SWB1 and SWB6. In addition, we assume that all non-directional OC protection relays use time delay settings t_s from the base case shown in Table 4.2.

In the case of fault between SWB1 and SWB2 (Figure 4.16) there is no problem in detecting and selectively isolating the fault from the main grid side by CB1.2, because the fault current seen by CB1.2 becomes higher $I_{kmax} = 15$ kA (Figure 4.17) than 13.5 kA (Figure 4.11) in the

Table 4.3 Scenario A2: Status of CBs and directional OC protection settings t_s

		CB1	CB2	CB1.1	CB1.2	CB2.1	CB2.2	CB3.1	CB3.2
Upper feeder		1	1	1	0	0	1	1	1
	→	0.75	0.55	0.55	0.4	0.4	0.3	0.3	0.2
	←		0.05	0.05	0.1	0.1	0.15	0.15	0.2

			CB3	CB4.1	CB4.2	CB5.1	CB5.2	CB6.1	CB6.2
Lower feeder			1	1	1	1	1	1	1
	→		0.55	0.55	0.4	0.4	0.3	0.3	0.2
	←		0.05	0.05	0.1	0.1	0.15	0.15	0.2

	CB1.3	CB1.4	CB1.5	CB2.3	CB2.4	CB2.5	CB3.3	CB3.4	CB3.5
DER + load	0	1	1	0	1	1	0	1	1

	CB4.3	CB4.4	CB4.5	CB5.3	CB5.4	CB5.5	CB6.3	CB6.4	CB6.5
DER + load	0	1	1	0	1	1	0	1	1

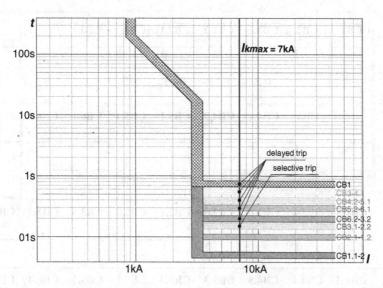

Figure 4.14 Base case trip curves and a tripping sequence in scenario A2 with directional OC protection

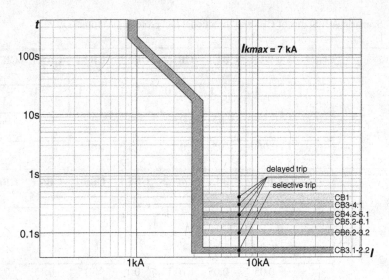

Figure 4.15 Modified base case trip curves (t_s settings) and a tripping sequence in scenario A2 with non-directional OC protection

base case due to a contribution from the synchronous DER at SWB1. The fault current supplied by the second DER at SWB6 and seen by CB2.1 is 2 kA (Figure 4.17). It can only activate the L part of the relay's trip curve with the expected tripping time delay of 40 s. Therefore, CB2.1 is opened by the follow-me function of CB1.2 and isolates the fault from the LV feeder side in $t_s = 150$ ms (if using a directional OC protection then $t_s = 400$ ms, see Table 4.3).

Table 4.4 Scenario A2: Status of CBs and modified non-directional OC protection settings t_s

	CB1	CB2	CB1.1	CB1.2	CB2.1	CB2.2	CB3.1	CB3.2	
Upper feeder	1	1	1	0	0	1	1	1	
	0.4	0.2	0.2	0.15	0.15	0.05	0.05	0.1	
		CB3	CB4.1	CB4.2	CB5.1	CB5.2	CB6.1	CB6.2	
Lower feeder		1	1	1	1	0	1	1	
		0.3	0.3	0.2	0.2	0.15	0.15	0.1	
DER + load	CB1.3	CB1.4	CB1.5	CB2.3	CB2.4	CB2.5	CB3.3	CB3.4	CB3.5
	0	1	1	0	1	1	0	1	1
DER + load	CB4.3	CB4.4	CB4.5	CB5.3	CB5.4	CB5.5	CB6.3	CB6.4	CB6.5
	0	1	1	0	1	1	0	1	1

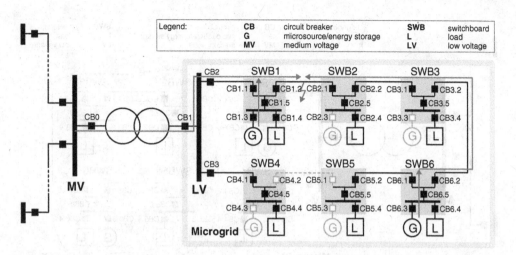

Figure 4.16 Scenario B1: the microgrid with synchronous DERs connected to the MV distribution grid

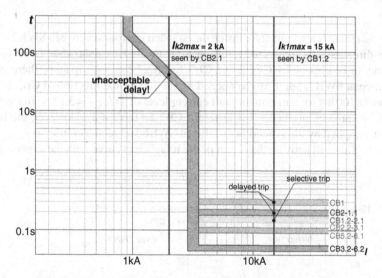

Figure 4.17 Base case trip curves and a tripping sequence in scenario B1 with directional OC protection

The main concern is $t_s \geq 150$ ms set for the OC relay in CB1.2, which may affect the stability of the synchronous DER with a small inertia at SWB1. A preferred solution is based on the adaptive directional interlock. The time delay t_s is set at 50 ms for all OC relays in the microgrid. Then blocking signals are sent in the correct directions, which prevents an unnecessary disconnection of DERs and healthy parts of the microgrid.

Next we assume that, after an isolation of the first fault, the island which includes SWB2, 3, 5 and 6 is formed as shown in Figure 4.18. The synchronous DER in SWB6 is switched to a frequency and voltage control mode, and additionally each load in the island is reduced from 100 A to 50 A.

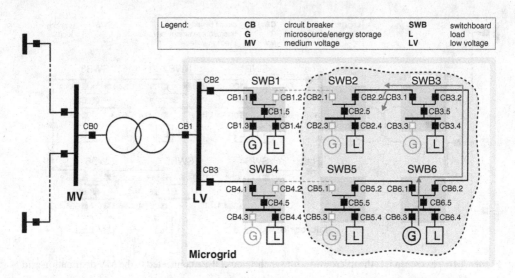

Figure 4.18 Scenario B2: the islanded microgrid with the synchronous DER in SWB6

Assume that there is a second fault inside the islanded microgrid between SWB2 and SWB3 and all non-directional OC relays use t_s settings from the base case shown in Table 4.2. Ideally, the fault should be cleared by CB2.2 and CB3.1. CB2.2 can not trip since there is no fault current source at SWB2, but it can be opened by the follow-me function of CB3.1. The t_s of CB3.1 is set at 100 ms for a minimum fault current level of $4*I_{n\,CB} = 3.2$ kA. In case of using directional OC protection, then $t_s = 150$ ms for CB3.1 (Table 4.3). However, the maximum fault current supplied by the synchronous DER at SWB6 and seen by CB3.1 is $I_{kmax} = 2.4$ kA (Figure 4.19). This will activate the L part of the relay's trip curve with the expected tripping

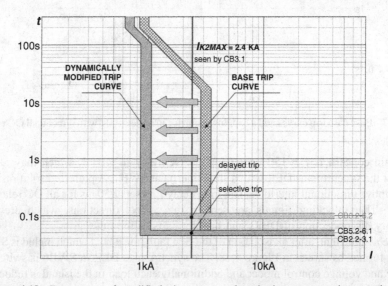

Figure 4.19 Base case and modified trip curves and a tripping sequence in scenario B2

time delay of 25 s. During this time, the DER at SWB6 will be disconnected by its out-of-step protection.

In order to guarantee fast fault isolation in the islanded mode where the main grid does not contribute to the fault, the trip curve must be pushed to the left dynamically, depending on the microgrid topology and a number of connected DERs. The modified trip curves for scenario B2 are illustrated in Figure 4.19.

Protection settings for all CBs in the island are calculated during the offline fault analysis and shown in Table 4.5, as recorded in the event and action tables.

Another protection alternative is based on the adaptive directional interlock. The tripping time is set at 50 ms for all CBs inside the island, and the minimum short-circuit current has to be dynamically modified (reduced) depending on the type and number of connected DER units.

A more elaborate example of multilevel microgrid and centralized adaptive protection system is provided in the Appendix 4.A.1.

Table 4.5 Scenario B2: Status of CBs and modified OC protection settings I_{kmin} and t_s

		CB1	CB2	CB1.1	CB1.2	CB2.1	CB2.2	CB3.1	CB3.2
Upper feeder		1	1	1	0	0	1	1	1
	I_{kmin}	3.2	3.2	3.2	3.2	3.2	1.2	1.2	1.2
	t_s	0.3	0.2	0.2	0.15	0.15	0.05	0.05	0.1
		CB3	CB4.1	CB4.2	CB5.1	CB5.2	CB6.1	CB6.2	
Lower feeder			1	1	0	0	1	1	0
	I_{kmin}		3.2	3.2	3.2	3.2	1.2	1.2	1.2
	t_s		0.2	0.2	0.15	0.15	0.05	0.05	0.1
	CB1.3	CB1.4	CB1.5	CB2.3	CB2.4	CB2.5	CB3.3	CB3.4	CB3.5
DER + load	1	1	1	0	1	1	0	1	1
	CB4.3	CB4.4	CB4.5	CB5.3	CB5.4	CB5.5	CB6.3	CB6.4	CB6.5
DER + load	0	1	1	0	1	1	1	1	1

4.3.5 Adaptive Protection System Based on Real-Time Calculated Settings

An alternative centralized approach to the one based on predefined settings is to recalculate protection settings in real-time operation, immediately upon changes in the microgrid topology (network reconfiguration or DER connections). This protection scheme can be implemented as a multifunctional intelligent digital relay (MIDR). When a fault occurs, the routine implemented in the MIDR generates selective tripping signals, which are sent to respective circuit breakers. The MIDR allows continuous real-time measurement and monitoring of the analog and digital signals originating from equipment and network, respectively. Integration of state estimation routines into a configuration of the protection system, which monitors the DER operating state and sets up a corresponding data exchange with the protection system is also possible [13]. Thus, new network operating conditions can be assessed, functioning of the protection system can be analyzed and, if required, protection settings can be adapted. Figure 4.20 illustrates the algorithm of the developed adaptive protection relay. The scheme consists of two blocks: real-time and non-real time.

The real-time block analyzes the actual microgrid state, acquired by continuous measurement of microgrid parameters, and detects disturbances due to adjusted tripping characteristics of protection devices. When a tripping condition is detected, a tripping signal to the respective circuit breaker is generated by the MIDR. The non-real-time block uses the prediction data of the availability of DERs in order to review the selectivity of the tripping characteristics for each new operating condition and to adapt them accordingly, if selectivity is not suitably provided any longer. If the adaptation is successful and the boundary conditions are not violated, the tripping characteristics of respective relays will be matched. If there is no possible solution without violating the boundary conditions, a signal will be generated that forbids the acceptance of the operation predicted by the (decentralized) energy management system (DEMS/EMS).

The proposed adaptive protection concept for microgrids with protection settings calculated online was tested and proved in the 10/0.4 kV simulator in the laboratory of Fraunhofer IWES (Figure 4.21). The simulator is equipped with photovoltaic systems, combined heat and power units, diesel generators, models of wind power plants and different types of inverters.

In the test setup, the MIDR comprises the functionalities of both DEMS and MV/LV station controller. For each of the three feeders (transformers T-1 to T-3), there is a separate database server, realized by means of MySQL servers. In the databases, data about the interconnected distributed energy resources (i.e. available short-circuit power provided by DER units) and the actual microgrid configuration is stored. The actual interconnection states and short-circuit power of DER units are stored in the respective databases and continuously communicated to the MIDR. Communication between the MIDR and the database servers can be realized via PLC or Ethernet. Based on the requested operational data (short-circuit power of DER in the different grid sections) and the actual short-circuit power available from the main grid, the MIDR calculates possible short-circuit currents and adapts the protection settings, where required. According to the protection characteristics and the measurement signals supplied by the current and voltage transformers, the MIDR decides whether there is a fault or not, and sends tripping signals to respective circuit breakers, if required. In this concept, the MIDR plays the role of the centralized protection controller which limits its application to a single substation case. For remote circuit breakers, communication can become a bottleneck for fast

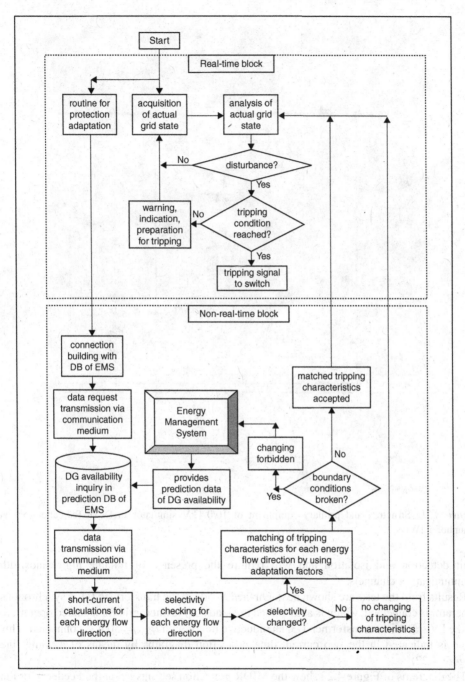

Figure 4.20 Simplified flow chart of adaptive protection system concept

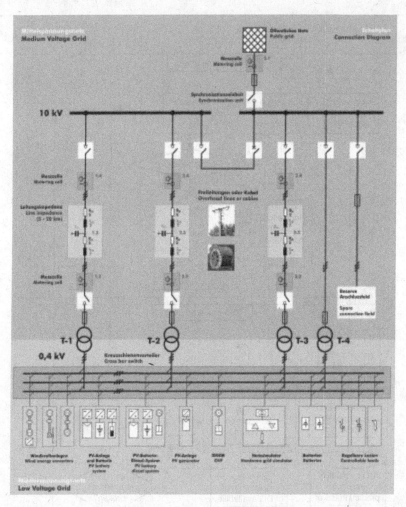

Figure 4.21 Structure and primary equipment of 10/0.4 kV simulator in DeMoTec laboratory at Fraunhofer IWES

fault detection and isolation and will require the presence of reliable high-bandwidth communication channels.

Results from the tests are shown next. On the LV side of the transformer T-1, a synchronous generator is connected, which is able to feed the loads of the microgrid for islanded operation. At the LV side of the transformer T-3, an inductive motor with $I_{rated} = 50$ A is connected. This value is indicated in the two current-time protection diagrams by means of a white line (Figure 4.22).

The diagrams of Figure 4.22 show the MIDR protection settings $I(t)$ at the Feeder 3 for the grid connected (left) and the islanded (right) operating conditions. The bottom line shows the simplified motor startup characteristic, the top line indicates the short-circuit characteristic (damage curve) of the LV cable installed between the motor and the transformer T-3 and the middle curve is the tripping characteristic of Feeder 3, which controls the operation of the

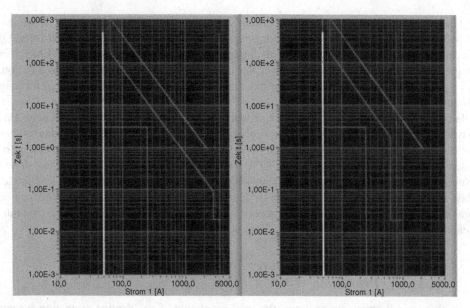

Figure 4.22 I–t protection diagrams for the feeder 3 (T-3) for the grid connected (left) and the islanded (right) operation modes

circuit breaker at the LV side of transformer T-3. The tripping curve is set between the motor startup characteristic and the damage curve of the LV cable. Its high set element ($I\gg$) is set to 0.8 times the maximum available short-circuit current.

For the grid connected operation the maximum available short-circuit current at the Feeder 3 is 3700 A, which is indicated by the vertical line. Therefore, $I\gg$ (right part of the middle line) is set to $0.8 \times 3700\,A = 2960\,A$ for the grid-connected operation. For the isolated operation (full short-circuit power is provided only by local DER) the maximum available short-circuit current at Feeder 3 is reduced more than four times to 800 A. Therefore, $I\gg$ is now set to $0.8 \times 800\,A = 640\,A$. The MIDR automatically adapts the protection settings in accordance with the conditions of the microgrid. When a fault occurs in the microgrid, the middle line is exceeded and tripping signals are generated by the MIDR in order to trip the respective circuit breaker(s). The MIDR then behaves like a conventional protection relay which controls more than one circuit breaker.

Similarly to the example above, during transition from grid-connected to islanded operation, the protection settings will be adapted for a change in the interconnected DER (i.e. a change in short-circuit contribution provided by DER), since some DERs will be left upstream of the point of connection. The MIDR continuously monitors the DER availability (relevant data available from database servers) and the main grid availability (if applicable) and estimates the short-circuit currents for every short-circuit flow direction. The vertical line of the above diagrams, which indicates the estimated short-circuit current, is shifted accordingly and the curve adapts to the new conditions. The speed of protection adaptability is limited by the probability of fault occurrence during adaptation. During the tests carried out, the time duration for adaptation has been estimated to be in the range of a few seconds.

The suggested adaptive protection concept does not consider the function of automatic reclosing, which is usually employed for temporary faults occurring on overhead lines. Of special relevance to this concept is the data transfer between adaptive relays and EMS over long distances. Communication via power line carrier (PLC) and LAN is a preferred option. For a practical implementation, wireless radio communication should be considered for long distance communication.

For the implementation of the adaptive network protection concept in distribution networks, existing communication systems and protocols can be applied. The IEC 61850 standard allows for real-time communication by means of periodically sent telegrams. For a sudden change of state at the sending unit, a telegram (GOOSE message = Generic Object Oriented Substation Event) will be send with a high repetition rate within a few milliseconds after the change of state. Thus, changes in the microgrid state can be detected rapidly. The communication link can be established by means of common interfaces (RS-485, fiber optic etc.) and protocols. Communication architectures and protocols are discussed in Section 4.3.6.

4.3.6   Communication Architectures and Protocols for Adaptive Protection

As explained in the previous sections, the implementation of an adaptive protection system requires that protection intelligent electronic devices (IEDs) communicate between themselves or with a central controller. By these communications, IEDs can provide and receive information about the actual topology of the microgrid, about the status of generators, energy storage devices and loads, and commands to execute an action (e.g. change of the active setting group). The application of pre-calculated protection settings, can be deployed via a centralized or a decentralized communications architecture:

- A centralized architecture is the most conventional and common communication scheme. There is a central controller that takes decisions about protection IEDs which have different settings for different operating conditions. This architecture is supported by many communication protocols, the most common ones within the electric energy sector being: Modbus, DNP3, IEC 60870-5-101/104 and IEC 61850. The centralized architecture can be deployed with serial communications, bus communications, over PLC or through an Ethernet network.
- A decentralized architecture is less conventional. Within this architecture it is not necessary to employ a central controller that coordinates protection settings, because intelligence is distributed between IEDs themselves. Each IED, with the information received from other IEDs, acts autonomously in order to change its active setting group. This architecture is only feasible when the communication protocol allows it to establish a direct communication between the IEDs, substituting direct hardwired connections between the relays. For standard protocols, today the industry's focus is on IEC 61850. The decentralized architecture needs to be implemented over a bus or Ethernet network, although with a suitable bandwidth it can be implemented with future 4G wireless networks or over PLC.

The most significant advantage of a centralized architecture is that local protection devices do not take decisions and therefore they are simpler. The central controller processes data and makes decisions based on the data received from the local devices. For example, to implement an adaptive protection system, the programmable logic is solely run on the central controller. It receives information about the status of all switching devices (feeder, DERs and loads) from local protection devices, such as IEDs. After the decision is taken, the central controller sends setting switching commands to the corresponding IEDs. The main disadvantage of a centralized architecture is its full dependence on the central device. Failure of this central element means total loss of the adaptive protection system, so redundancy is needed, which means additional cost.

The biggest advantage of the decentralized architecture is that it does not depend on one central controller. Its biggest disadvantage is the requirement for more advanced local protection devices, since each local device will need to run dedicated programmable logic for autonomous adaptation of settings depending on microgrid operating conditions.

Regarding the communication protocols that enable a protection system to become adaptive, the IEC 61850 standard receives most interest. Its main disadvantage is that it requires an Ethernet network. Ethernet is a common technology for automation at HV and MV (the previous section covered adaptive protection based on IEC 61850), but today, it still means extra cost and technical complexity in LV installations. Moreover, it poses additional complexity for the maintenance personnel, who are more familiar with hardwired systems.

The main advantages of IEC 61850 are:

- can be applied to every type of electrical installation and is able to functionally cover any application,
- guaranteed interoperability between devices from different vendors,
- standardized data models,
- simplified engineering process and IED configuration,
- scalable solutions.

For adaptive protection application, this standard has communication services for the usual operations within electrical installations, such as change of settings and setting groups, reporting of information, commands and so on. One remarkable service is the GOOSE (generic object oriented substation event) service, already mentioned in Section 4.3.5. This service makes direct information exchange between IEDs possible, accepting any type of data (single signals, double signals, measurements etc.), guaranteeing that it is transmitted in less than 3 ms (faster than a cable between digital outputs and inputs), and is always available because it is retransmitted indefinitely. It comprises a multi-master system that offers the highest reliability. Obviously, the application of GOOSE messages for communication between IEDs within the microgrid, provides a valuable tool allowing them to adapt their configurations to any operational scenario. Each IED can receive data from all the others very rapidly, so the reconfiguration process is very fast.

Figure 4.23 shows an example of how the GOOSE messaging can be used in a microgrid. A protection IED at the LV side of the distribution transformer in bay 1C issues a signal to block the trip of protection IEDs within the microgrid in bays 1A, 1B, 1D and at the IED immediately on the LV side of the transformer. Protection IEDs on those bays need only subscribe to that GOOSE message in order to effectuate the blocking scheme. GOOSE messages can be used to inform other IEDs about topological changes in a microgrid (e.g. transition from grid connected mode to islanded mode), to implement protection schemes

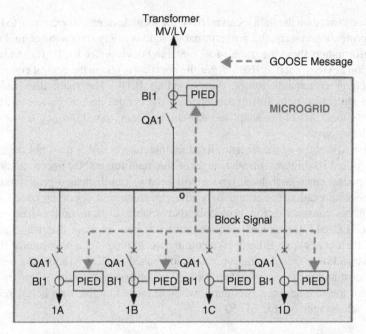

Figure 4.23 The blocking scheme based on GOOSE for faults in the microgrid

(like the one in Figure 4.23), to execute commands (e.g. open/close circuit breakers), to change active setting group and so on.

An adaptive protection system, such as one designed according to the IEC 61850 standard, requires higher investment than a traditional protection system for LV based on fuses and hardwired interlock. However, microgrid operation will not be feasible without an adaptive protection system. Perhaps one of the main concerns in considering the IEC 61850 solution is the cost of the network, especially the cost of devices that were not necessary in traditional LV networks, the "switches". Switches enable the organizing of the traffic within the network, taking into consideration several parameters, such as the priority of a message, or managing VLANs in order not to mix different traffics. Switches are necessary, but it is possible to use cheaper architectures by means of integrating these switches inside IEDs. The implementation of an Ethernet network as a single ring of IEDs with integrated switches is a very good option in order to reduce the cost of a high-performance microgrid.

4.4 Fault Current Source for Effective Protection in Islanded Operation

As discussed in Section 4.1, one of the major problems in microgrid protection is the low contribution to faults by DERs that have power electronics (PE) interfaces. This section analyzes the installation of an independent system that supplies a fault current in the microgrid dominated by DERs with PE interfaces, for example, PV panels. The proposed system, designed as a fault-current source (FCS), is based on power electronics but uses a simpler and thus less costly technology than is normal in inverters in this power range. An FCS would consist of the following subsystems:

1. an energy storage device capable of supplying the active power for the fault current for a specified duration – the amount of power and energy required depends on the anticipated rating of the fault current source. The most economical way to provide a large amount of power for a very short time (small energy) is in the form of an ultra-capacitor; these are characterized by low standby losses and are maintenance free. They have a long life expectancy, if used predominantly in idle mode.

2. a switch or power electronic circuit capable of releasing the power from the fault current source into the LV network; if an ultra-capacitor is used as a storage device, a PE circuit is needed to convert the DC voltage across the capacitor into an alternating current. For this application, the optimum design uses much silicon and moderate cooling, which implies using separate insulated-gate bipolar transistor (IGBT) modules on a simple heat sink. In LV distribution networks, most faults behind the customer meter are single phase to earth/ neutral faults. Therefore the inverter must provide a neutral connection and preferably act independently for each phase. This approach minimizes the annoyance to users of the other two phases during a single-phase fault.

3. a detector which monitors the LV network and signals the condition of a fault, triggering the power circuit of the switch to supply the fault current – the detection system includes the intelligence of the fault current source. It determines when the source has to be activated and it injects current into the microgrid. In order to be able to operate an FCS as a plug-and-play device, this circuit should act only on locally measured information. As the FCS is connected in parallel to the LV network, the obvious detection criterion is the network voltage at the interconnection point of the FCS. It is assumed that the FCS is connected to the microgrid at the point where the MV/LV step-down transformer would normally be connected. As soon as the busbar voltage drops to a value below its statutory minimum, the FCS can assume that something irregular has happened in the microgrid network. This could be a fault, but it could also be insufficient generation to meet the demand. Whatever the cause, the FCS response will not harm the microgrid operation. If the voltage drop is caused by a transient overload – for example caused by the starting current of a large motor – the FCS will reduce this overload. In any case, because the FCS is designed to cease operating after a predetermined number of seconds, the other sources in the microgrid will eventually supply the necessary loads again. In principle, the FCS consists of one energy storage unit connected to three inverter phases that serve the three network phases independently. Therefore, the triggering system will monitor each phase independently and trigger only the inverter phase corresponding to the faulted phase. This approach is valid for single-phase to neutral faults and for three-phase faults, because in both situations the phase-to-neutral voltage drops to a low value. However, on a phase-to-phase fault, the voltages of the two affected phases may simply phase-shift towards each other. If that happens, all phase-to-neutral voltages remain more or less within specification, but the three-phase system has a large negative-sequence component. This can be addressed by also measuring the phase-to-phase voltages and triggering both affected inverter phases if the voltage between the corresponding phases drops below a set threshold.

4. a recharging system that restores the state of charge of the energy storage device to its maximum level after a fault current has been delivered – if an inverter is built with an ultra-capacitor in its DC link, it is not possible to connect it directly to the LV network unless it is charged to a certain threshold voltage. A discharged ultra-capacitor would behave almost like a short circuit. Because of its large capacitance, the inrush current when connecting the inverter to the network would last for many seconds and probably destroy the inverter's diodes. For this reason, the system must be built according to a "slow charge, fast discharge" philosophy. A suitable implementation uses a transformer to avoid high inrush currents. The inverter is used to boost the output voltage of the transformer so that the ultra-capacitor stack is charged to its full voltage.

The operating principles of an FCS are illustrated in Figure 4.24. Its power circuit remains idle during normal operation of the network (1). Whenever a fault occurs, the network voltage drops (2), the FCS is activated (3) and it attempts to restore the original network voltage, thereby injecting a fault current into the network. The FCS has finite impedance and, as a consequence, the voltage could be lower than the nominal voltage of the network. The current is high enough to cause a fuse or circuit breaker to clear the fault. Subsequently, the FCS maintains the original voltage and frequency (4), to enable inverters that had turned off to resynchronize and reconnect to the network. After some time, typically 5 seconds, the FCS can turn off because the full load is now supplied by the local sources in the microgrid.

In most LV networks, there is a statutory requirement to detect and remove a low-impedance fault within a few seconds, for example 5 seconds. Therefore after 5 seconds the FCS can assume that the fault has been cleared and that it is justifiable to cease operation. However, this should not happen instantaneously, because some sources in the microgrid may have switched off as a consequence of the voltage dip during the fault and need some time to reconnect. Therefore, at the end of the 5 second interval, the FCS will remain operating, while gradually increasing its virtual impedance, and then cease operation. Once triggered, the FCS acts like a low-impedance voltage source. The most sensible location for an FCS in a microgrid is therefore the main LV distribution busbar of the distribution step-down transformer, if the microgrid is normally supplied from an MV network. In this way, a

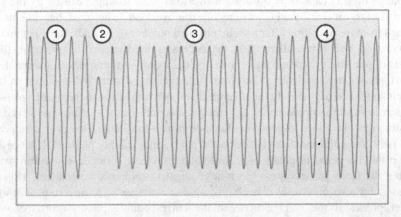

Figure 4.24 Typical voltage waveforms illustrating the operation of an FCS

classical radial topology is possible with straightforward graded protection settings along individual feeders. In order to avoid a misoperation of the protections in the network, it is recommended to always install DERs, whose fault current contribution is much smaller than the fault level created by the FCS at the first protection device upstream of the DER. If the decision to install a DER is beyond the authority of the network operator, the available DERs have to be taken into account when setting the protections.

In order to demonstrate the operating principle of FCS, a single-phase prototype with the following specifications has been used:

Network voltage	230 V single-phase
Maximum fault current	200 A rms
Energy storage technology	ultra-capacitors
Capacitance of UC stack	0.44 F
ESR of UC stack	2.3 Ω
Maximum voltage of UC stack	840 VDC

The FCS would require a minimum voltage of ~585 VDC to be able to restore the network voltage to a value within statutory limits. This means that the energy effectively available to be injected into the network is 80 kJ. The effective series resistance (ESR) of the ultra-capacitor limits the amount of power that can be delivered by the FCS. If a voltage drop of 20% of the maximum voltage is accepted, the maximum DC discharging current is 73 A and the maximum DC power to be delivered is 49 kW. Technical details of the construction of the FCS and its control logic are provided in Appendix 4.A.2.

Failure of a network to trigger a thermal fuse can be considered as a failure to deliver the required amount of energy as soon as a fault occurs. For this reason, the fault detection algorithm uses the thermal value of the voltage (the squared value) as a variable to monitor. The squared value of a 50 Hz sinusoidal voltage varies sinusoidally at a frequency of 100 Hz. An accurate measurement of the thermal value therefore involves averaging over an integer number of half-periods. The quickest way to observe changes in the thermal value is by using a moving average of the squared voltage over one half-period. The speed of response of the detection depends on the threshold that is applied. A reasonable upper value of the detection threshold corresponds to 80% of the rated voltage. In this way, the system will never respond to small transients even if the stationary grid voltage is on its lower statutory limit of 90%. The 80% corresponds to a threshold 64% (0.8^2) for the moving average of the squared voltage. A collapse of the voltage to 10% is detected within 2 ms, as shown in Figure 4.25.

Laboratory tests from the operation of the prototype in parallel to a utility power supply with additional high series impedance are presented next. The circuit diagram is shown in Figure 4.26. Some of the tests have been done using a standard D type thermal fuse, as depicted in the diagram. Other tests have been performed using an automatic circuit breaker with thermal/magnetic operation. The 7 Ω impedance in series with the supply ensures that a short circuit, as indicated in the diagram, will lead to a fault current of ~30 A, which is too low to trigger either the fuse or the circuit breaker within 5 seconds.

Figure 4.27 depicts the voltage across the FCS and the current through the short circuit. The fuse at the end consumer premises blows after 180 ms. In the particular configuration of this prototype, the FCS is able to blow thermal fuses up to 35 A. If larger fuses need to be blown, the required energy will scale more or less linearly with the required current. As energy

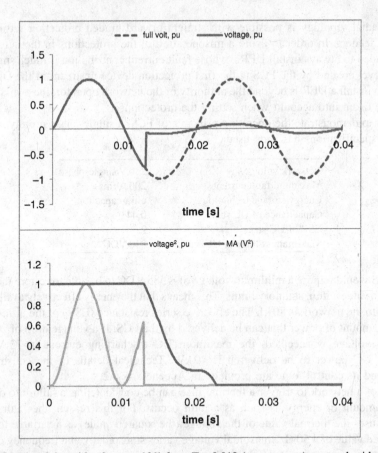

Figure 4.25 Dip of the grid voltage to 10% from T = 0.013 (upper traces), squared grid voltage and moving average of V^2 over one half-period (lower traces)

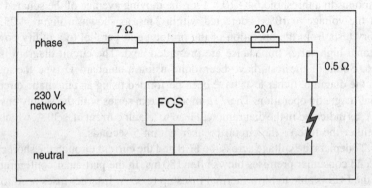

Figure 4.26 General test setup, demonstrating the functionality of the FCS

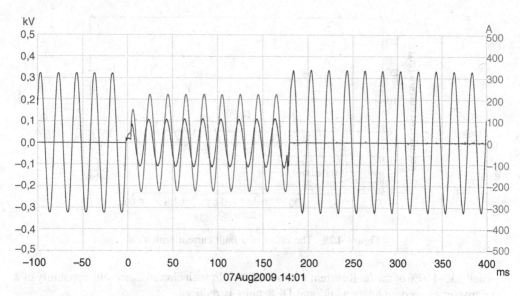

Figure 4.27 Voltage across the FCS (zero before and after event) and current through the short circuit (larger sinusoid during the event). Protection device: fuse DII 20 A

content is the main cost driver of an FCS, an FCS capable to blow, for example, a 200 A fuse would be six times bigger and six times more expensive than the prototype.

4.5 Fault Current Limitation in Microgrids

Distribution utilities calculate fault current levels during the network planning stage to ensure that fault levels remain within the design limits of various grid components. These calculations are based on knowledge about connected generating units and rotating equipment at customer sites. In today's distribution networks, the presence of DERs provides an additional contribution to the fault level, and the embedded nature of the DERs makes the fault current calculations more complex, as they should take into account the consequences of operational combinations to a degree not required when all generation is coming from the transmission network [7]. Even if the fault current contribution from a single DER is not large, the aggregated contribution of many DERs can raise the fault current levels beyond the design limits of various equipment components (e.g. circuit breakers, cables, busbars etc.). When the fault level design limits are exceeded, there is a risk of damage to, and failure of, the equipment, with consequent risks of injury to personnel and interruption of supply under short-circuit fault conditions. In this case, a costly upgrade of equipment may be needed.

The fault level contribution from a DER is mainly determined by its type and its coupling method to the grid. Today, DERs connected to LV microgrids are commonly PV generators and micro and mini CHPs. Existing LV microgrids can accept a large number of PV generators. The impact of PV covering 100% of the microgrid load would add less than 5% to the commonly used fault levels. However, for large numbers of domestic mini CHPs equipped with synchronous generators, fault currents may increase considerably, especially in urban areas with low fault-level headroom availability (a fault current level with DER can

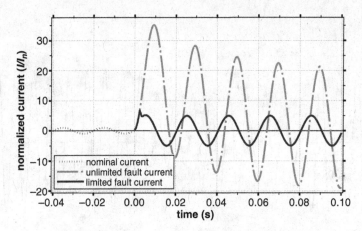

Figure 4.28 The effect of a fault current limitation

reach 130–150% of the fault current without DERs). In such circumstances the capability of a microgrid to accommodate additional DER units is reduced.

The duty of a fault current limiter (FCL) is to limit the excessive currents, in the case of a fault. The current-limiting effect is illustrated in Figure 4.28. In case of fault, the fault current increases with a certain rate of rise, depending on the circuit parameters and the phase angle at fault initiation. Usually, the FCL has to react and limit the current before the first peak of the fault-current waveform is reached (<5 ms). Figure 4.29 demonstrates available and emerging technical solutions for current limitation below equipment design limits [14]. These solutions fall in two groups: passive and active.

Passive solutions limit the fault current by increasing its current path impedance at nominal and post-fault conditions. Therefore, it generates losses and a voltage drop under nominal conditions. Active devices exhibit a highly nonlinear behavior and quickly increase the impedance during the fault. Their main disadvantage is the need to replace them after each operation and a careful adjustment of the protective relay settings to maintain selectivity [15]. There are also LV circuit breakers available with a built-in current limiting functionality [11]. The current limitation is done by a sufficient voltage build-up inside the breaker. The current path inside a CB is constructed in such a way that magnetic forces quickly move the arc into a set of metallic plates, making the arc split into a number of smaller arcs, each causing a voltage drop

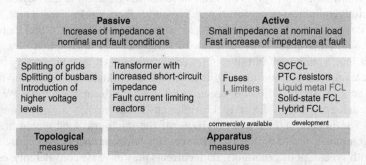

Figure 4.29 Alternative methods of fault current limitation

of 20–30 V and limiting the fault current [16]. However, the current limiting circuit breakers only limit the current in case of actuation and always interrupt. Selectivity is therefore limited.

CIGRE WG A3.10 [14] provides a list of requirements for the ideal FCL:

- negligible impedance under nominal operating conditions – that is, negligible resistive and reactive losses, no related needs for cooling and no voltage drop;
- fast response – the first peak of the short-circuit current (<5 ms) must be limited in order to limit the magnetic forces on the primary equipment components;
- multiple operations – the FCL device has to recover automatically, with very short recovery time in order to be able to respond to multiple re-closure cycles and in order not to require a field trip for service personnel to replace components after operation;
- selectivity – the let-through current should be selectable, either by the design of the FCL device or by a configurable setting. It should not limit motor start currents and should not respond to transients or capacitor switching currents. The current for coordination with protective devices has to be provided, so that existing protection concepts do not need to be modified;
- high reliability – the FCL must correctly operate under any fault magnitude and any fault phase condition. Correct response must reliably occur after a long duration without fault events as well as in cases of consecutive multiple faults;
- compact size, long lifetime, maintenance free and low cost.

In Table 4.6, compliance with the requirements of the ideal FCL is given for different technologies.

A device called an I_s-limiter combining a fast switch and a current limiting fuse in parallel (Figure 4.30) has been developed [17]. Under normal operation the current flows via the low impedance switch. Upon detection of a fault current by the electronic control circuit, the fast switch is triggered and the current is commutated to the fuse. By using a small charge for

Table 4.6 Compliance of different technologies with the requirements for the ideal FCL

Technology	Requirements					
	Nominal impedance	Response time	Repeatability	Selectivity	Reliability	Costs, size
CL transformer	V drop	fast	given	ok	high	high, small
CL reactor	V drop	fast	given	ok	high	med, med
Resonant circuit	ok	fast	given	ok	high	high, large
CL fuse	ok	fast	replace	limited	high	low, small
I_s-limiter	ok	fast	replace	limited	high	high, small
CL circuit breaker	ok	fast	ok	limited	high	low, small
PTC resistor	ok	medium	cooling	limited	medium	low, small
Superconducting FCL	ok	fast	cooling	ok	not proven	high, med.
Liquid metal FCL	ok	medium	ok	ok	not proven	med, small
Driven-arc FCL	ok	fast	ok	limited	not proven	med, med
Semiconductor based FCL	ok	fast	ok	limited	not proven	high, med.
Hybrid FCL	ok	fast	cooling	limited	not proven	med, med.

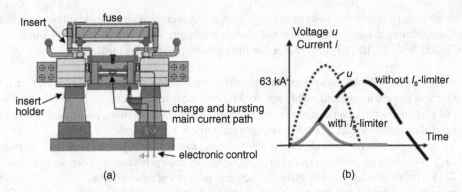

Figure 4.30 Is-limiter with insert (a) and its current limiting effect (b)

interrupting the main current path, the I_s-limiter is able to interrupt the fault current within 0.5 ms after receiving the tripping signal. Due to the electronic control the tripping conditions can selectively be chosen. The I_s-limiter is a single-shot device, and the switch/fuse insert needs to be replaced after actuation. I_s-limiters are available for up to 40.5 kV and for rated currents of several kA.

If the current is not allowed to be completely interrupted by the Is-limiter, a parallel combination of an Is-limiter with a reactor can be used. Upon tripping of the Is-limiter the current is commutated on the reactor, which limits the fault current. The elimination of the permanent resistive copper losses of the reactor makes the Is-limiter very cost-effective.

4.6 Conclusions

Effective protection is fundamental for the successful deployment of microgrids. The main challenges are posed by the interconnected DERs resulting in varying operating conditions depending on their status, reduced fault contribution by those interfaced by power electronic inverters and occasionally increased fault levels. This chapter focuses on automatic adaptive protection, in order to modify protection settings according to the microgrid configuration. Techniques to increase the amount of fault current level helped by a dedicated device, and possible use of fault current limitation are also discussed.

Appendices:

4.A.1 A Centralized Adaptive Protection System for an MV/LV Microgrid

In this Appendix a more elaborate application of the adaptive protection system based on pre-calculated settings is provided.

In the microgrid of Figure 4.31 the islanded operation is supported at two voltage levels: the complete MV microgrid can be disconnected from the utility grid and alternatively the LV microgrid can be disconnected from the MV microgrid.

At the MV level, the system includes a single point of common coupling to the utility grid and a number of CBs controlled by IEDs equipped with a directional over-current protection function, shown in Figure 4.31 as PTOC. Each IED supports up to six independent setting groups and is fully compliant with IEC-61850 standard. All IEDs installed at the MV level are integrated into the centralized IEC-61850 automation system with a rugged controller

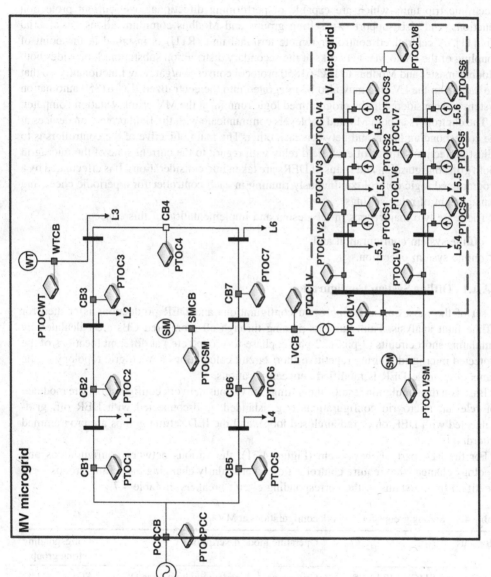

Figure 4.31 A centralized adaptive protection system for the multilevel microgrid

installed at the primary substation (point of common coupling) serving as the core for making the control decisions, as well as providing data and event logging, reporting and human–machine interface (HMI) functionality.

At the LV level, the system consists of a number of LV circuit breakers also equipped with electronic trip units which are capable of performing directional over-current protection functions, and also support two setting groups and Modbus communications via a serial bus [11]. A centralized controller remote terminal unit (RTU) is installed at the point of coupling of the LV to MV networks at the secondary distribution substation. It provides both Modbus master and Modbus to IEC-61850 protocol conversion/gateway functionality, so that the data from the LV side may also be integrated into the centralized IEC-61850 automation system and considered in the programmed logic running at the MV primary station computer.

The controllers at both MV and LV levels communicate with the field protection devices at the corresponding levels and between each other. The main objective of the controllers is to adjust protection settings of each field relay with regard to the current state of the microgrid (both grid configuration and status of DERs are taken into consideration). It is effectuated by a programmed logic control continuously running in each controller for a periodic check and management of relay settings.

There are two major steps in the design and implementation of this concept:

- offline system configuration and
- online system operation.

4.A.1.1 Offline System Configuration

A set of the most common microgrid configurations and DER productions is created for offline fault analysis. Fault currents passing through all monitored CBs are calculated by simulating short circuits (3-phase, 2-phase, phase-to-ground etc.) at different locations of the protected microgrid. During repetitive short-circuit calculations a microgrid topology or the status of a single DER is modified between iterations.

Based on the simulation results, the settings for various network configurations are modeled for relevant microgrid configurations (e.g. islanded, grid-connected with DER off, grid-connected with DER on or radial/closed loop), and the IED setting groups are programmed accordingly.

For the MV part of the system (Figure 4.31), the various network configurations and topology changes that require control action – particularly changing the setting groups – are identified by the status of the corresponding circuit breakers in Table 4.7.

Table 4.7 Setting groups for network configurations at MV level

CB status	Operating mode description	Corresponding setting group
PCCCB – on, SMCB, CB4 – off	Grid connected mode; radial; no large DG	SG1
PCCCB – on, SMCB – on, CB4 – off	Grid connected mode; radial; Large DG contribution	SG2
PCCB – off, SMCB. WTCB – on	Islanded	SG3
All CBs on	Grid connected mode; closed loop; large DER contribution	SG4

If a breaker is not listed in the table, its status is unimportant for the control logic. Also, for convenience and to simplify the monitoring from the SCADA operator side, it is decided that all IEDs must adjust their setting groups to exactly the same active setting group number, even though in some cases the actual settings may be identical for several (all) setting groups. This is done so that situations are avoided where some of the IEDs, for example, are switched to setting group 2, while the others are still at setting group 1.

Electronic protection devices used at the LV level are only capable of keeping in memory two sets of setting groups. Therefore, we use the first setting group for the grid-connected mode and the second group is activated during islanded operation. However, we anticipate that future LV protection devices will have more than two setting groups which may further improve sensitivity, selectivity and speed of protection, especially during islanded operation of the microgrid with different types and sizes of DERs.

The next step after selection of setting groups is configuration of communications:

- IEC61850 for MV and
- Modbus-RTU for LV.

4.A.1.2   IEC 61850 Configuration

The overall offline engineering process for the proposed adaptive protection system is shown in Figure 4.32.

First, a system study is required to determine the IED settings for four setting groups in configurations outlined in Table 4.7, and the IEDs are configured remotely or via a local HMI.

Then the IEC 61850 automation system is created and the reporting is organized as follows. IEC 61850-7-4 Edition 2 defines the standard logical nodes and data objects related to the IED setting groups and circuit breaker position. IEC 61850-7-2 Edition 2 describes the mechanism and the associated abstract communication service interface (ACSI) for both reporting and control of the IED setting groups, which is called the setting group control block (SGCB). The ActSG service parameter of the setting group control block ACSI model can be used to read and write the new setting group value if a change is required due to certain changes in the microgrid topology.

Another standard data attribute used for monitoring the network configuration changes in the system is the circuit breaker position. It is a part of XCBR logical node and can also be included in a dataset. In terms of the IEC 61850 common data classes' definition, the circuit breaker position is described as controllable double point (DPC) with the status value data attribute stVal, which can take four values representing the four different possible states of the circuit breaker: 0 – intermediate, 1 – off, 2 – on, 3 – bad.

The ability to use the standard attributes is an important advantage of IEC 61850; it increases interoperability and allows for easy integration of the IEDs from multiple vendors into the automation system, although true plug-and-play functionality is not yet achieved. For instance, even though the circuit breaker position data attribute is standardized, it may be located in different logical devices for IEDs coming from different vendors.

For the purpose of changing the setting groups based on changes of microgrid topology, the status of the active setting group and the current circuit breaker position must be reported by

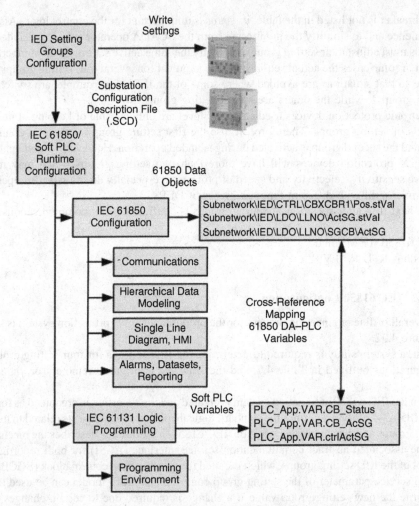

Figure 4.32 System configuration for a dynamic protection settings management at MV level

the IEDs to the substation computer which runs the logic and takes the control decisions. In this case, the central controller (substation computer) acts as a client in the IEC 61850 client/server environment. Therefore these data objects must be either included in a dataset to which the substation computer will subscribe, or be configured to be sent as a GOOSE message. It has been found that the standard IEC 61850 dataset event reporting of the circuit breakers position provides acceptable performance. This method also saves bandwidth and simplifies sending the data across hybrid networks.

The substation computer is represented in the IEC 61850 SCL hierarchy as an IED client device with just one logical node for the HMI. Then the buffered report control blocks are configured for all actual IEDs to send the reports to the substation computer client spontaneously, that is, when a change of the circuit breaker position occurs.

To provide additional dependability, integrity reports may also be enabled, and the integrity polls issued by the client to retrieve and confirm the status of the circuit breaker

Table 4.8 IEC 61850 standard data objects

Object reference	Attribute name	Type	Read/write	Description
LD0.LLN0.SGCB	ActSG	INT8U	W	write active setting group
LD0.LLN0.ActSG	stVal	INT32	R	read active setting group status value
CTRL.CBXCBR1.Pos	stVal	DPC	R	status of the circuit breaker

periodically rather than spontaneously. If the integrity polls are used, it is recommended to create a separate dataset just for the active setting group and circuit breaker status to reduce bandwidth consumption, since the integrity poll will build a report containing all values of the referenced dataset, not just the ones that have been changed since the last poll.

The IEC 61850 data objects and attributes used to monitor the position of the circuit breaker and the active setting group status and control are summarized in Table 4.8.

4.A.1.3 Logic Processor Configuration

The logic processor is the brain of the adaptive protection system. It uses the IEC 61131-3 compatible languages to implement the logic used to extract the information received from the IEC 61850 communications infrastructure, perform the analysis and, if necessary, execute the control action by sending a control command back to the IEC 61850 automation system.

Table 4.9 illustrates the cross-mapping of the IEC 61850 objects to the PLC server variables for one of the IEDs.

The rest of the IEDs are mapped in a similar manner, and the PLC variables are organized as arrays. The cross-reference mapping includes the direction which defines whether an IEC 61850 attribute is being monitored or controlled by the logic processor.

4.A.1.4 Communications Integration

The system includes the IEC 61850 OPC server, which integrates all the IEC 61850 compatible IEDs with the associated data objects. It also includes the logic processor OPC server which is responsible for monitoring and control. Additionally, the system contains an instance of Modbus TCP OPC server for the integration of the LV part of the network shown in Figure 4.31 through a TCP connection with LV RTU, as well as the DNP LAN slave

Table 4.9 IEC 61850 to logic processor cross references

From Server Name	OPC Server Path	PLC Server Path	Direction
IEC61850 OPC server	WA1\CB1\CTRL\CBXCBR1 \Pos\stVal	PLC_LOCAL.SettingsChange. GVL.bCB_stVal[1]	→
IEC61850 OPC server	WA1\CB1\LD0\LLN0 \ActSG	PLC_LOCAL.SettingsChange. GVL.iActSg_stVal[1]	→
IEC61850 OPC server	WA1\CB1\LD0\LLN0\SGCB \ActSG	PLC_LOCAL.SettingsChange. GVL.iActSg_ctlVal[1]	←

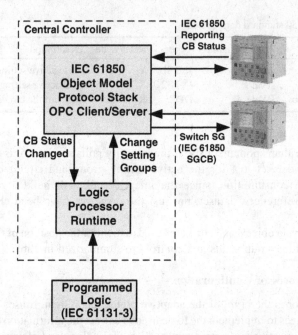

Figure 4.33   Real-time communications architecture

OPC client used to map the data from the IEC 61850/Modbus microgrid automation system to the upstream DNP3 SCADA master.

4.A.1.5   Online Data Exchange and System Operation

The schematic diagram for the online operation of the system is shown in Figure 4.33.

The real-time data exchange is organized as follows. The IEDs at the MV level use IEC 61850 communications; the substation computer runs the IEC 61850 OPC server, in order to receive the report data from the relays. It also executes the monitoring and control logic executed within the runtime logic processor engine. The IEC 61850 data attributes are mapped to the logic processor variables at the offline configuration stage as described in the previous section. Therefore, the logic processor application has fast access to the data received from the IEC 61850 side through the OPC data access mechanism and can detect the system events that will trigger the execution of the adaptive protection module and dictate the need for the control action. It must be noted that there is a delay associated with transferring the data between process signals and logic variables. The shortest possible data transfer cycle is 50 ms.

In situations where a setting's change is required, the logic processor runtime again uses the OPC DA mechanism to connect to the IEC 61850 environment and execute the command to change setting groups. The setting group change will only occur when the circuit breaker is in either on or off position; it will be inhibited if the status of the circuit breaker is undetermined (bad) or intermediate, which means that a circuit breaker is possibly in the middle of the reclosing sequence. The settings change will also be inhibited if the quality attribute of the IEC 61850 data object is bad. After the command to adjust the setting groups has been sent to the IEDs in the next cycle of its operation, the logic will read the ActSG status to make sure that the command was indeed successful.

The system also includes an additional OPC server instance, which can be used to map the microgrid communications data to the distribution network control center via SCADA. A protocol conversion may be executed at this step, if necessary; for example, IEC 61850 objects may be mapped to DNP3.

The logic developed with the IEC 61131-3 compatible tools can be used without changes at both MV and LV.

4.A.2 Description of the Prototype FCS

The circuit diagram of the prototype FCS is shown in Figure 4.34.

The ultra-capacitor stack UC is composed of 320 cells of 140 F connected in series. The cells are mounted on printed circuit boards each containing 16 cells plus voltage balancing resistors. Capacitors C1A and C1B are electrolytic capacitors acting as a DC-link capacitor in parallel to the IGBT stack. Electrolytic capacitors have a much lower ESR than the ultra-capacitor stack, and serve to reduce the ripple on the DC voltage during operation of the IGBTs. Also, the series connection of two capacitors is used to create a neutral connection to the network. The IGBT stack is formed by IGBTs V1 and V2 and by freewheeling diodes D1 and D2 in a half-bridge configuration. Both during charging and when operating during a fault, the half-bridge generates a pulse width modulated (PWM) sinusoidal voltage. The modulation frequency is 6 kHz. Inductor Lf and capacitor Cf together form the output filter of the FCS. The output filter removes the high-frequency components from the voltage generated by the half-bridge and converts the PWM voltage into a sinusoidal voltage. Moreover, inductor Lf acts as the output impedance of the FCS. The value of Lf is 3 mH, resulting in a physical output impedance of $\sim$1 Ω at 50 Hz.

Transformer Tc and contactors K2 form the pre-charging circuit of the FCS. If the ultra-capacitor stack is fully discharged, contactors K2 will be closed to start the charging sequence. The secondary no-load voltage of transformer Tc is 80 V. The impedance of the transformer and the inductor Lf together limit the rms AC input current to 50 A. The primary input current

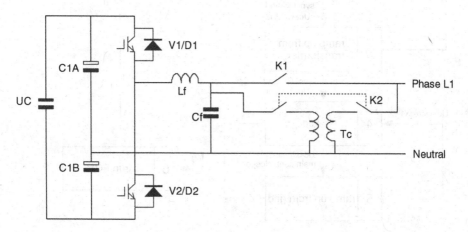

Figure 4.34 Circuit diagram of the prototype FCS

of the transformer is then 18 A, which is an acceptable load for the network. The voltage Uc across the ultra-capacitor increases to 150 VDC within 4 seconds. As soon as $Uc > 150$ VDC, the half-bridge starts switching. The PWM generation is controlled in such a way that the current in inductor Lf has an rms value of 10 A and that the current is in phase with the network voltage. Thus the reactive component is zero. In this way, the FCS takes a constant charging power of 800 W from the network. Subtracting the losses in transformer, inductor and IGBTs, the power delivered to the ultra-capacitor stack is ~500 W. This part of the charging phase ends as soon as $Uc > 600$ VDC, which takes ~2.5 minutes in this configuration. This DC voltage is high enough to justify direct connection of the half-bridge to the network. Consequently, contactors K2 are opened, and contactor K1 is closed. The control parameters of the PWM generation are changed to represent the new topology. Now the current in inductor Lf is controlled to an rms value of 4 A. The ultra-capacitor is further charged until $Uc = 840$ VDC (during the tests reported here, this maximum was set to 820 VDC).

For safety reasons the FCS may be required to store enough energy to clear more than one fault, because another fault could occur during the recharging period. On the other hand, the full energy content is only needed to trigger the biggest fuse on the most remote fault. If the clearing time is shorter than worst-case, only a small fraction of the energy content is absorbed during a single fault.

The control system operates according to the state transition diagram of Figure 4.35. The system has five operating states and two administrative states, which are described below. If the system was restarted or rebooted while there was still voltage in the DC buffer, the system

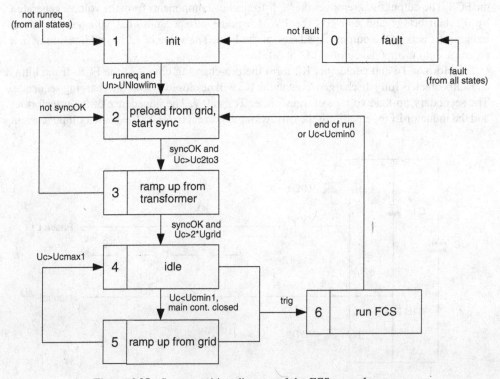

Figure 4.35 State transition diagram of the FCS control system

would automatically proceed through the diagram until it reached the state which is most appropriate for the actual DC voltage. During a reboot, all contactors are opened and the IGBTs are switched to a passive state so that possibly destructive transients cannot occur.

- *state 0* – Whenever a hardware-related fault has occurred, the system enters state 0. The system will remain in state 0 until all fault conditions have been removed and an appropriate delay time has elapsed. Within state 0, the fault reset logic has its own state machine.
- *state 1* – When the FCS is started or the software is rebooted, the system enters state 1. In state 1, all variables are initialized. The FCS can be enabled/disabled from the user interface.
- *state 2* – Contactors K2 are closed so that the ultra-capacitor is charged via the pre-charging transformer, and pre-charging continues until the DC voltage exceeds 2 to 3 times a threshold *Uc*, which is presently set to 150 VDC. When both the synchronization and the voltage threshold condition are fulfilled, the system will enter state 3.
- *state 3* – Contactors K2 remain closed, but now PWM on the inverter will start so that the DC voltage can be further ramped up. The half-bridge is operated in current-control mode. The measured current in inductor Lf is decomposed into an active (in-phase) component and a reactive (out-of-phase) component. These components are the inputs of two controllers, whose outputs are used to calculate the phase and amplitude of the sinewave generated by the PWM. The reference for the reactive part is zero, the reference for the active part is set to $-10\,A$. The minus sign is used because the positive polarity of the current is defined as flowing from the FCS into the grid. This state is finished when the DC voltage *Uc* is high enough for the inverter to be connected directly to the network, which is the case when $Uc > 2*Ugrid$, that is, twice the peak value of the grid voltage. The system then enters state 4.
- *state 4* – This is the idle state, in which the system is ready for operation but remains idle, to minimize losses. Because both the ultra-capacitor stack and the electrolytic capacitors are provided with voltage sharing resistors, the DC voltage will decrease gradually and a regular top-up is needed to retain the voltage. This recharge is done in state 5 and is initiated if $Uc < UCmin1$. Threshold parameter *UCmin1* is presently set to 780 V. This also means that if the system enters state 4 from state 3 (condition of transition: $Uc > 2* Ugrid$, which is in the order of 600 V), it will immediately move on to state 5. State 4 is purposely used as an intermediate state. A hardware interlock ensures that contactor K1 can only be closed after contactors K2 have opened and the system is only allowed to enter state 5 after the closing of contactor K1 has been confirmed.
- *state 5* – The operation of state 5 is very similar to that of state 3. However, the inverter is now connected to the grid directly, instead of via the transformer, which means that the settings of the PWM are different and that the two current controllers use different control parameters and reference values. In state 5, the reactive current controller has a reference value of a few amps capacitive, so that the inverter will supply the capacitive current flowing into filter capacitor Cf. In this way, the power factor of the input current remains close to 1. In the prototype, the reference value for the active current controller is set to 4 A.
- *state 6* – From both states 4 and 5, the system is allowed to enter state 6. This transition occurs when the fault detection logic has detected a probable fault in the network. During the transition to state 6, the sinewave generator for the PWM is detached from the

phase-looked loop (PLL) and remains running with a fixed frequency value. The voltage dip associated with the fault may have caused the PLL to observe a frequency step. Therefore the measured value of the frequency is saved in a FIFO buffer in 100 ms. If a fault is detected, the PWM generation continues on the basis of a 100-ms-old value of the frequency, which generally will be a reliable pre-fault value. In state 6, the current controllers are inactive and the PWM uses a fixed amplitude for the output voltage, equal to the pre-fault value of *Ugrid*. In addition to that, the PWM calculation includes a virtual impedance which can virtually add both inductive and resistive components to the output inductor. With the lapse of time, the virtual impedance will increase, which makes it possible for the other generators in the network to regain control of the voltage and phase of the network. After a preset time – now set to 5 seconds – the FCS will leave state 6 and enter state 2. Depending on the remaining DC voltage after this action, the system may subsequently proceed to state 3, 4 or 5. If during its action in state 6, the DC voltage drops too low, defined as $UC < UcminO$, the system will immediately enter state 2. The value of *UcminO* is presently set to 600 V.

References

1. Feero, W., Dawson, D., and Stevens, J. "Protection Issues of the Microgrid Concept". Available http://certs.lbl. gov/pdf/protection-mg.pdf.
2. Al-Nasseri, H., Redfern, M.A., and O'Gorman, R. (2005) "Protecting microgrid systems containing solid-state converter generation", Int. Conf. on Future Power Systems.
3. Vilathgamuwa, D.M., Loh, P.C., and Li, Y. (2006) Protection of microgrids during utility voltage sags. *IEEE T. Ind. Electron.*, **53** (5), 1427–1436.
4. Driesen, J., Vermeyen, P., and Belmans, R. (2007) Protection issues in microgrids with multiple distributed generation units, 4th Power Conversion Conf., Nagoya.
5. Nukkhajoei, H. and Lasseter, R.H. (2007) "Microgrid Protection", IEEE PES General Meeting.
6. EPRI (2006) "Overview of SEMI F47-0706" [online]. Available www.f47testing.com.
7. KEMA Limited (2005) "The contribution to distribution network fault levels from the connection of distributed generation".
8. Bower, W. and Ropp, M. (March 2002) "Evaluation of islanding detection methods for photovoltaic utility interactive power systems," IEA-PVPS, Online report T5-09. Available: http://www.oja-services.nl/iea-pvps/ products/download/rep5_09.pdf.
9. Jayawarna, N. (2004) "Fault current contribution from power converters", EU Microgrids project, Available www.microgrids.eu.
10. Rockefeller, G.D. *et al.* (1988) Adaptive Transmission Relaying Concepts for Improved Performance. *IEEE Transactions on Power Delivery*, **3** (4), 1446–1458.
11. ABB SACE SpA (2007), "Low voltage moulded case circuit breakers Tmax", Product brochure, Available http:// library.abb.com/global/scot/scot209.nsf/veritydisplay/117ea5b751d87f12c125734e002a25dc/$File/ 1SDC210015B0203.pdf.
12. ABB SACE SpA (2008), "DOC simulation tool", User Manual version 1.0.
13. Shustov, A. and Valov, B. (2006) Lösungswege der Integrationsprobleme dezentraler Energieerzeugungsanlagen in elektrischen Verteilungsnetzen, Vorträge des internationalen wissenschaftlich-technischen Seminars, Polytechnische Universität Tomsk, 10.11.04.
14. CIGRE WG (2003) A3.10 "Fault Current Limiters – Fault Current Limiters in Electrical Medium and High Voltage Systems", CIGRE Technical Brochure, No. 239.
15. CIGRE WG (2006) A3.16 "Guideline on the impacts of fault current limiting devices on protection systems", A3-06 (SC) 33 IWD.
16. Lindmayer, M., Marzahn, E., Mutzke, A. *et al.* (2004) "The process of arc-splitting between metal plates in low voltage arc chutes", 22nd Int. Conf. on Electrical Contacts, Seattle.
17. ABB Carol Emag (2008) "Is limiter", Product brochure, Available http://library.abb.com/global/scot/scot235.nsf/ veritydisplay/31656b3a0912262ac12574f700565d1e/$File/2243-08%20E_300dpi.pdf.

5

Operation of Multi-Microgrids

João Abel Peças Lopes, André Madureira, Nuno Gil and Fernanda Resende

5.1 Introduction

This chapter aims to study the management of distribution networks with an increased microgrid penetration, that is, corresponding to a situation where most of the low voltage (LV) networks turn into active microgrids. It is therefore assumed that the microgrid concept is extended, leading to the development of a new concept – the multi-microgrid [1]. A full exploitation of this concept involves the design of a new control architecture as well as the development of new management tools or the adaptation of existing distribution management systems (DMS) tools.

The novel concept of multi-microgrids corresponds to a high-level structure, formed at the medium voltage (MV) level, consisting of several LV microgrids and distributed generation (DG) units connected to adjacent MV feeders. For the purpose of grid control and management, microgrids, DG units and MV loads under active demand-side management control can be considered as active cells in this type of power system. In this new scenario, the possibility of having some MV responsive loads able to receive control requests under a load curtailment strategy is regarded as a way to ensure additional ancillary services.

A large number of LV networks with microsources and loads, that are no longer passive elements of the distribution grid, then need to be operated together in a coordinated way. Therefore, the system to be managed increases largely in complexity and dimension, requiring a completely new control and management architecture. This architecture is illustrated in Figure 5.1.

An effective management of this type of system requires the development of a hierarchical control architecture, where control will be exercised by an intermediate controller – the central autonomous management controller (CAMC) – to be installed at the MV bus level of a high voltage (HV)/MV substation, under the responsibility of the distribution system operator (DSO) and will be in charge of a multi-microgrid system. In this way, the complexity of the

Microgrids: Architectures and Control, First Edition. Edited by Nikos Hatziargyriou.
© 2014 John Wiley & Sons, Ltd. Published 2014 by John Wiley & Sons, Ltd.
Companion Website: www.wiley.com/go/hatziargyriou_microgrids

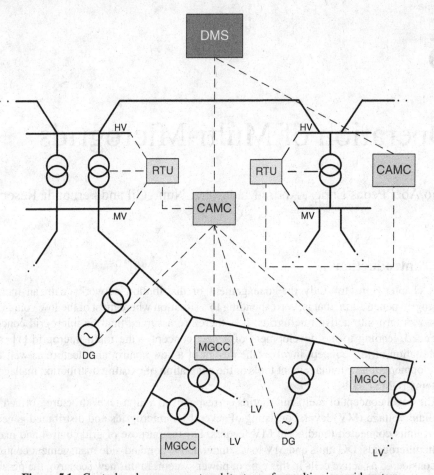

Figure 5.1 Control and management architecture of a multi-microgrid system

system can be shared among several smaller individual control agents, behaving like a small DMS that is able to tackle the scheduling problem of generating units (DG and microsources) and other control devices that are installed in the system in both normal and emergency conditions.

In this context, the deployment of a communications infrastructure as a means for achieving full observability of the distribution network is crucial. This can be achieved by exploiting a smart metering infrastructure, which will allow a coordinated and integrated management of individual elements at the LV level, such as microgrids (and corresponding microgenerators, loads and storage devices). DG units and loads participating in active demand-side management strategies, directly connected to the MV grid, may require a dedicated communication infrastructure. This local communication infrastructure will be the key driver for managing the distribution network in a more efficient way and consequently maximizing the integration of DG and microgeneration, especially renewable-based power sources.

5.2 Multi-Microgrid Control and Management Architecture

The development of the microgrid concept has led to the need for a detailed analysis of the interaction between the microgrid central controller (MGCC) and the central DMS (operated by the DSO), which is in charge of the whole distribution network, in both normal and emergency operating modes. This has already been addressed in Chapter 2, and several other publications also describe it [2,3].

However, as previously seen, the new concept of multi-microgrids is related to a higher-level structure, formed at the MV level, consisting of LV microgrids and DG units connected to adjacent MV feeders.

Nowadays, the DMS is wholly responsible for the supervision, control and management of the whole distribution system. In the future, in addition to this central DMS, there may be two additional management levels:

- *HV/MV substation level*, where a new management agent, the CAMC, will be installed as illustrated in Figure 5.1. – the CAMC will accommodate a set of local functionalities that are normally assigned to the DMS (as well as other new functionalities) and will be responsible for interfacing the DMS with lower-level controllers.
- *microgrid level*, where the MGCC to be housed in MV/LV substations will be responsible for managing the microgrid, including the control of the microsources and responsive loads – voltage monitoring in each LV grid will be performed using the microgrid communication infrastructure.

Similarly to what happens in the microgrid concept, in multi-microgrid systems, two modes of operation and operating state may be envisaged. These two possible operating modes are:

- *normal operating mode* – when the multi-microgrid system is operated interconnected to the main distribution grid.
- *emergency operating mode* – when the multi-microgrid system is operated in an autono- mous mode (islanded from the main power system) or, following a blackout, when the multi-microgrid system is contributing to service restoration by triggering a black start procedure.

The main issue when dealing with control strategies for multi-microgrid systems is the use of individual controllers, which should have a certain degree of autonomy and be able to communicate with each other in order to implement specific control actions. A partially decentralized scheme is justified by the tremendous increase in both dimension and complexity of the system so that the management of a multi-microgrid system requires the use of a more flexible control and management architecture.

Nevertheless, decision-making – even with some degree of decentralized control – should still adhere to a hierarchical structure. A central controller should collect data from multiple devices and be able to establish rules for low-rank individual controllers. These actions must be set by a high-level central controller (the DMS), which ought to delegate some tasks to other lower-level controllers (either the CAMC or the MGCC). This is because central management would not be effective owing to the large amount of data to be processed, and therefore would not allow autonomous management, namely during islanded mode of operation. The CAMC must then have the ability to communicate with other local controllers (such as MGCCs, or directly with DG sources or loads connected to the MV network), serving as an interface for the main DMS.

Consequently, the CAMC plays a key role in a multi-microgrid system as it will be responsible for the data acquisition process, for enabling the dialog with the DMS located upstream, for running specific network functionalities and for scheduling the different resources in the downstream network. In this context, existing DMS functionalities need to be upgraded due to the operational and technical changes that result from the adoption of the multi-microgrid concept, especially the introduction of the CAMC, and corresponding hierarchical control architecture.

The management of the multi-microgrid (MV network included) will be performed through the CAMC. This new controller will also have to deal with technical and commercial constraints and contracts, in order to manage the multi-microgrid both in HV grid-connected operating mode and in emergency operating mode, as previously seen. A set of functionalities to integrate the CAMC is shown in Figure 5.2.

It is important to stress that all these functionalities may not be available in all multi-microgrid systems. Their availability will depend very much on the type of network considered and on the characteristics of local DG units, as well as on the active load demand-side resources availability.

Furthermore, in multi-microgrid systems, microgrids – together with MV-grid connected DG units – can participate in the provision of ancillary services, such as coordinated voltage support, with a significant impact on the electrical distribution system operation and also regarding its local dynamic behavior.

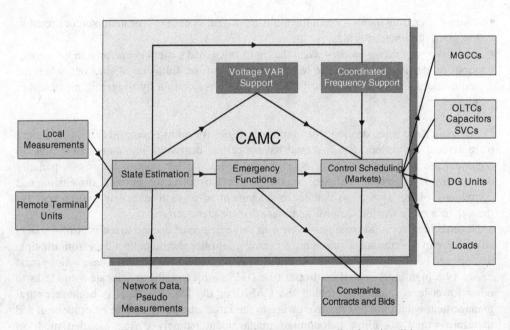

Figure 5.2 Central autonomous management controller functionalities

The need to study the dynamic behavior of a multi-microgrid, to evaluate the islanding feasibility, involves dealing with hundreds (or even thousands) of microgenerators, which may lead to a significant computational burden when developing these simulations. So dynamic equivalents for microgrids are of utmost importance for developing such studies.

Several key aspects of multi-microgrid operation, developed within the framework of the EU R&D project, More Microgrids [4,5], are addressed in the following sections, namely regarding some of the most important functionalities available at the CAMC level. Bearing in mind the relevance of some of these functionalities and the need to perform some of the key studies, such as dynamic behavior analysis, the following topics are addressed:

- coordinated voltage/var support for normal operation,
- coordinated frequency control for islanded operation,
- local black start – restoration of the MV grid following a blackout,
- definition of dynamic equivalents for microgrids.

5.3   Coordinated Voltage/var Support

5.3.1   Introduction

Large-scale integration of DERs (DG in particular) may pose serious operational problems for the DSO. One of the major concerns that arises is related to the voltage rise effect as a result of the massive presence of these generation units in the distribution system, namely in LV networks [6]. In these systems, conventional voltage regulation methods are not suited for tackling this problem effectively as they do not address variable power injections provided by DG units, so voltage control is an issue that must be dealt with efficiently.

Due to operational issues, most DSOs require that DGs operate at zero reactive power or at a fixed power factor. This limits the amount of DG installed capacity, in order to guarantee admissible voltage profiles in the worst-case scenario. Several approaches can be found in the scientific literature that address voltage control in distribution grids with the presence of DG. However, these proposals do not take into consideration coordinated operation at MV and LV levels and/or the specific characteristics of distribution networks.

Consequently, new approaches for voltage control must be developed, based on the full use of the resources available, especially DG and microgeneration units, taking into account the specific characteristics of distribution networks (in particular, LV grids that usually have very resistive lines). This is particularly important in the case of LV networks where both active and reactive power control is needed for an efficient voltage control scheme. In some extreme cases, if no local storage is available, it may even be necessary to spill some local generation in order to avoid voltage rise problems [7].

5.3.2   Mathematical Formulation

A hierarchical coordinated approach for voltage/var control in distribution systems with high penetration of DG can be formulated as a multi-objective optimization problem. The proposed formulation of this problem aims at minimizing active power losses and microgeneration curtailment, subject to a set of technical and operational constraints such as voltage and line loading levels [8].

The main control variables identified are: reactive power from MV-connected DG sources, on-load tap changing (OLTC) transformer tap settings and microgeneration curtailment. Consequently, this problem becomes a mixed integer, nonlinear optimization problem, since it deals with both continuous and discrete variables.

The objective function and the constraints used to solve this problem are presented below.

$$min(\alpha \cdot \sum P_{loss} + (1 - \alpha) \cdot \sum \mu G_{shed}) \tag{5.1}$$

subject to:

$$P_i^G - P_i^L = \sum_{k=1}^{N} V_i \cdot V_k \cdot (G_{ik} \cdot \cos \theta_{ik} + B_{ik} \cdot \sin \theta_{ik}) \tag{5.2}$$

$$Q_i^G - Q_i^L = \sum_{k=1}^{N} V_i \cdot V_k \cdot (G_{ik} \cdot \sin \theta_{ik} + B_{ik} \cdot \cos \theta_{ik}) \tag{5.3}$$

$$S_{ik} \leq S_{ik}^{max} \tag{5.4}$$

$$V_i^{min} \leq V_i \leq V_i^{max} \tag{5.5}$$

$$Q_i^{min} \leq Q_i^G \leq Q_i^{max} \tag{5.6}$$

$$t_i^{min} \leq t_i \leq t_i^{max} \tag{5.7}$$

Where

α is a decision parameter to be defined by the decision-maker, in this case the DSO

P_i^G, P_i^L are the active power generation from DG and microgeneration and the active power consumption at bus i, respectively

V_i is the voltage at bus i

G_{ik} is the real part of the element in the admittance matrix (Y-bus) corresponding to the ith row and kth column

B_{ik} is the imaginary part of the element in the Y-bus corresponding to the ith row and kth column

θ_{ik} is the difference in voltage angle between the ith and kth buses

Q_i^G, Q_i^L are the reactive power generation/consumption at bus i, respectively

S_{ik} is the apparent power flow in branch ik

S_{ik}^{max} is the maximum apparent power flows in branch ik

V_i is the voltage at bus i

V_i^{min}, V_i^{max} are the minimum and maximum voltage at bus i, respectively

Q_i^G is the reactive power generation at bus i

Q_i^{min}, Q_i^{max} are the minimum and maximum reactive power generation at bus i, respectively

t_i is the transformer tap, or capacitor step, position of the OLTC transformer or capacitor bank i

t_i^{min}, t_i^{max} are the minimum and maximum taps of OLTC transformer or capacitor bank i.

Considering the specificities of the voltage control problem, particularly the fact that it comprises both discrete variables (transformer taps) and continuous variables (active or reactive power generation), a meta-heuristic approach – evolutionary particle swarm optimization (EPSO) – was used to find the solution to the problem. EPSO is a hybrid method based on two already well-established meta-heuristic optimization techniques: evolutionary strategies [9] and particle swarm optimization [10]. Several applications of EPSO to power system operation can be found in [11,12].

5.3.3 Proposed Approach

In order to develop an efficient coordinated method for voltage support in distribution grids, involving the MV and LV levels, the specific characteristics of these MV and LV networks must be considered. Concerning the MV network, a traditional power flow routine may be used to assess the impact of DG and microgeneration. However, for an LV system comprising single-phase loads and microgeneration units that cause phase imbalance, traditional power flow routines are not suitable. For this type of systems, a three-phase power flow must be employed in order to evaluate the expected impacts in the operation of the LV grid.

Furthermore, the main problem when dealing with optimizing distribution network operation is the dimension of the distribution system. Given the size of both MV and LV real distribution networks, a full representation of an MV network (including all LV feeders located downstream) is impractical. Since the dimension of the system may be huge, considering that an MV including the downstream LV networks can have several thousand buses, it becomes practically unfeasible to develop an algorithm using a full model representation of the MV and LV levels, able to operate in a real-time management environment.

Therefore, a decoupled approach was adopted that allows modeling in detail only those LV networks considered as critical in terms of microgeneration penetration levels. In this approach, from the MV point of view, each active microgrid is considered as a single bus with an equivalent generator (corresponding to the sum of all microgeneration) and equivalent load (corresponding to the sum of all LV loads), and passive LV networks are modeled as a non-controllable load as can be observed in Figure 5.3.

On the other hand, the effects of the control actions must be assessed at the LV level, especially in the case of "active" LV networks. This means that a detailed representation of these networks should be used in order to know the behavior of the main operation parameters, such as voltage profiles. This is extremely relevant since voltage profiles in LV networks with large-scale microgeneration integration tend to be high. The decoupled approach followed in this work allows modeling in detail only those LV networks considered as critical in terms of microgeneration penetration levels.

As a result, in order to speed up the integrated control algorithm, an artificial neural network (ANN) model able to emulate the behavior of the active microgrid was adopted. This option enables the use of the optimization tool employed in real-time operation, by reducing the long simulation times that are required in order to calculate consecutive LV power flows. In fact, the ANN can be regarded as an equivalent model, able to reproduce the steady-state behavior of the LV network, thus allowing the rapid evaluation of voltage profiles and other variables. Using the ANN, the computational burden – and consequently the computational time – required to run the application can be signifi-cantly reduced.

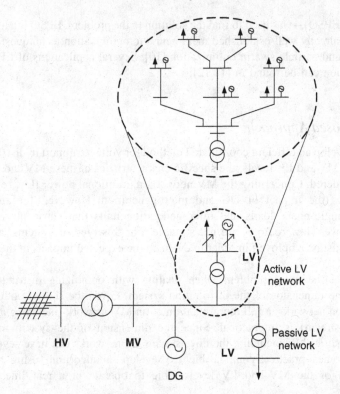

Figure 5.3 Decoupled modeling approach used for MV and LV networks

5.3.4 Microgrid Steady-State Equivalents

As previously mentioned, an ANN model able to emulate the behavior of an active microgrid for steady-state analysis has been developed. Therefore, each active microgrid can be replaced by an equivalent ANN model in the global optimization procedure as illustrated in Figure 5.4.

In this case, ANN models for each feeder of each LV network were developed, since it was observed that feeders may include different amounts of microgeneration and therefore the

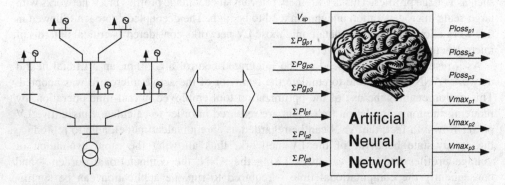

Figure 5.4 Neural network model for emulation of a microgrid in steady-state analysis

cause for an overvoltage may be originated from one single feeder with excessive micro-generation integration.

In Figure 5.4

V_{sp} is the voltage at the MV/LV substation

ΣPg_{pi} is the total active power generation in phase i

ΣPl_{pi} is the total active power load in phase i

$Ploss_{pi}$ is the total active power losses in phase i

$Vmax_{pi}$ is the maximum voltage value in phase i

In order to generate the dataset corresponding to the inputs chosen to train the ANN, a large number of three-phase power flows must be computed, considering different combinations of the inputs (i.e. several values for the voltage reference at the MV/LV transformer, for the power produced by each microgenerator and for the load), in order to calculate the active power losses and assess the voltage profiles in each of these scenarios.

For this purpose, a three-phase power flow routine was developed in order to enable the analysis of the steady-state behavior of LV networks, considering their specific character-istics: phase imbalances due to the connection of single-phase loads and single-phase microgeneration units exploiting different technologies.

The power flow solution method used is based on the one presented in [13] and is used for three-phase, four-wire radial distribution networks relying on a backward–forward technique, which is extremely fast to reach convergence.

From the tests that were performed, it was seen that the ANN with best performance was a two-layer feed-forward ANN with sigmoid hidden neurons and linear output neurons. Still, it must be pointed out that in case there is a change in topology of the "active" LV network, a new ANN must be computed. This means that it is necessary to update the data and generate a new dataset, followed by a retraining of an updated ANN, which may be considered as a drawback of this method.

In order to efficiently use the approach developed, it is necessary to automate the procedure of retraining the ANNs. First of all, the main changes that trigger the need for retraining the ANNs must be identified. These changes are mostly changes in the topology of the networks due to the inclusion (or removal) of additional DGs, microgeneration units or loads, the set-up (or upgrade) of a line or transformer and the expansion of the LV grid.

5.3.5 Development of the Tool

The algorithm developed for voltage control is designed to be used as an online function available for the DSO. This algorithm is intended to be integrated as a software module that will be housed in the CAMC.

Exploiting the communication infrastructure available in multi-microgrids, the CAMC will be able to collect critical information from the several devices in the network, namely data from load and RES forecast modules and generation dispatch, in order to assess voltage profiles, branch overload levels and active power losses. Using this information, the voltage control algorithm is run and, after it has successfully terminated, produces a set of commands in the form of setpoints. These commands are then sent to the several devices such as DG units, microgeneration units (via the corresponding MGCC) and OLTC transformers. This procedure is intended to run periodically and in an automated way, under the supervision of the DSO.

The topology and structure of the MV and LV distribution systems is also periodically updated in order to provide the various modules with data on the current status of the grid, including all devices such as DG units, microgenerators, loads and switches.

5.3.6 Main Results

The algorithm was tested using real Portuguese MV and LV distribution networks, considering expected future scenarios for the integration of DG and microgeneration units at the MV and LV level, respectively. These networks are presented and described in [7].

It was observed that the developed control approach is able to maintain secure operating conditions without the violation of technical constraints, namely in terms of voltage profiles. Figure 5.5 shows the daily profile of active power losses before and after running the voltage control algorithm.

Figure 5.6 shows the tap values at the OLTC transformer. According to the modeling of the transformer with taps on the secondary side, tap values above 1 raise the voltage on the secondary of the transformer. Consequently, lower tap values are used when voltage profiles are typically high and vice versa. As can be observed, the constraint used for limiting the number of switching actions of the OLTC transformer meant that, in consecutive one-hour periods, only one tap change from one period to the other was required.

Furthermore, the contribution of a MV-connected DG unit for voltage support is presented in Figure 5.7. The DG units were used to supply reactive power (positive values) or to absorb reactive power (negative values), according to the operation scenario requirements. It can be observed that the DG unit tends to absorb reactive power during daytime, when generation profiles are high, and voltage values are above the admissible limits. Conversely, during nighttime, the DG unit supplies reactive power in order to raise voltage profiles and help in reducing active power losses.

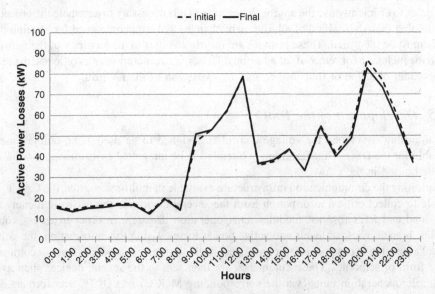

Figure 5.5 Total active power losses in the MV network

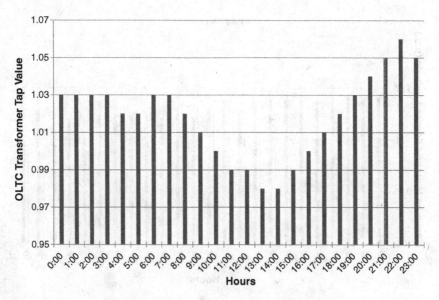

Figure 5.6   Tap changes in OLTC transformer

For the case analyzed, without voltage control, voltage values in some feeders were above the admissible range of +5% owing to the PV-based microgeneration, which is generating near its peak capacity. In this case, the PV panels generate power at peak capacity around 13:00, which is outside the peak demand hours and in some cases forces excess generation through the MV/LV transformer, causing reverse power flows.

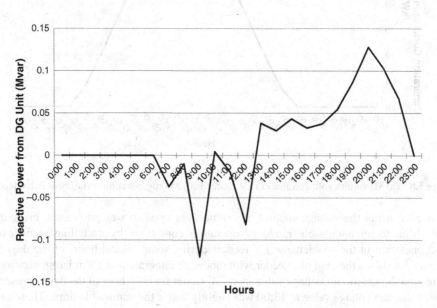

Figure 5.7   Reactive power from MV-connected DG unit

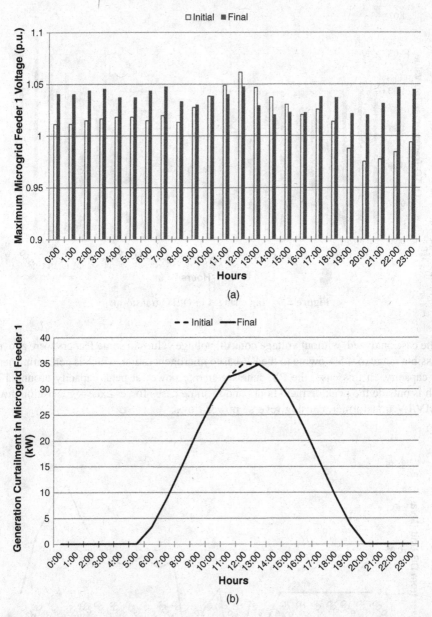

Figure 5.8 (a) Maximum voltage values in LV feeder 1 (b) Microgeneration curtailment in LV feeder 1

However, using the voltage support algorithm developed, it was possible to bring these values back to an admissible range of operating conditions by curtailing some excess microgeneration in the problematic LV feeders during some critical hours of the day.

Figure 5.8 shows the case of a feeder with moderate integration of PV microgeneration. In Figure 5.8a the maximum voltage value in an LV feeder for each hour of the day is presented. It shows that the voltage value at 13:00 was slightly above the admissible limit. Therefore, it

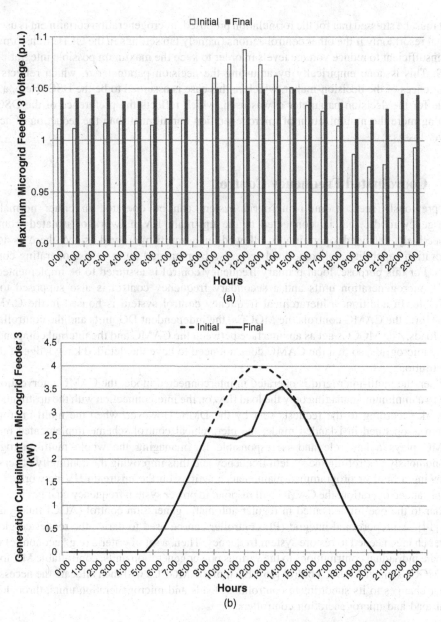

Figure 5.9 (a) Maximum voltage values in LV feeder 3 (b) Microgeneration curtailment in LV feeder 3

was necessary to curtail a small amount of PV microgeneration in order to bring the voltage back into the admissible range (Figure 5.8b).

Figure 5.9 presents a case with large integration of PV microgeneration. In this case, it was necessary to curtail PV microgeneration at various times of the day in order to ensure that there were no voltage violations.

It must be stressed that for the formulation proposed, microgeneration curtailment is used as a final resort, only if the other control actions (namely tap settings in the OLTC transformers) are insufficient to reduce voltage levels in order to keep the maximum possible integration of RES. This is done empirically by adjusting the decision parameter α, which reflects the preferences of the decision-maker, which in this case is assumed to be the DSO. Here, a low value for the decision parameter α was used, which reflects the preference of the DSO in valuing more the minimization of microgeneration curtailment than the reduction of active power losses.

5.4   Coordinated Frequency Control

As previously seen, a multi-microgrid system can be operated in either normal or emergency mode, that is, connected to the upstream HV network or isolated from it, respectively. Islanded operation is allowed in a multi-microgrid, since there is local generation capability and some loads can also be managed during these operating conditions. For this purpose, local primary frequency control is assumed to be implemented in some microgeneration units and a secondary frequency control is also supposed to be available. In addition, a hierarchical frequency control system is housed in the CAMC. Therefore, the CAMC controls the MGCCs, the independent DG units and the controllable MV loads. The MGCCs act as an interface between the CAMC and the internal components of the microgrids, so that the CAMC does not need to have the detailed knowledge of their constitution.

When the multi-microgrid is operated in interconnected mode, the CAMC intervention is kept to a minimum, managing only the load flow on the interconnection with the upstream HV network according to the requests sent by the DMS. However, when the multi-microgrid system is operated in islanded mode, the hierarchical control scheme implemented in the CAMC plays a key role and is responsible for managing the whole multi-microgrid, autonomously controlling the system frequency and thus improving the continuity of service following a fault or programmed maintenance actions on the upstream HV network.

In islanded operation, the CAMC will respond to power system frequency changes in a way similar to the one implemented in regular automatic generation control (AGC) functionalities [14]. A proportional-integral (PI) controller can be used to derive the requested global power change needed to restore system frequency. Then, a set of criteria (e.g., economic) will allocate individual contributions to the various power generation units, controllable MV loads and MGCCs under CAMC control [15]. Each of the MGCCs will then allocate the necessary power changes to its subordinate controllable loads and microgeneration units, through the local load and microgeneration controllers.

5.4.1   Hierarchical Control Overview

The hierarchical control system can be represented by the block diagram in Figure 5.10. When dealing with a single autonomous multi-microgrid (a single MV network) only control levels 2 and 3 need to have their function studied or simulated. The need to analyze any functions related to the DMS only arises while dealing with several MV networks at once, with more than one supervised by this DMS.

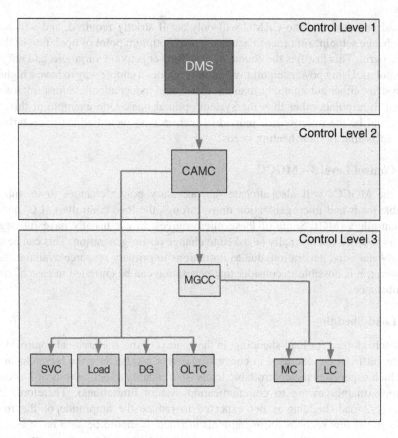

Figure 5.10 Hierarchical control scheme of a multi-microgrid system

5.4.1.1 Control Level 2 – CAMC

As mentioned earlier, the commands needed to modify power generation and loads originate in the CAMC. These commands are sent to MGCCs, to local DG units and also to controllable MV loads. MGCCs act as an interface between the CAMC and the internal active components of the microgrids, so that the CAMC doesn't need to have the details of each microgrid's constitution.

While the MV multi-microgrid is connected to the upstream HV network, the CAMC limits its autonomous intervention to a minimum. However, in islanded operation, the CAMC will step in and respond to power system frequency changes in a way similar to that implemented in regular AGC functionalities. A PI controller is then used to derive the requested global power change needed to restore system frequency. Then, an economic allocation algorithm will allocate contributions for this power change to the various power generation units, controllable MV loads and MGCCs under CAMC control, but only if they are willing, at that point in time, to participate in frequency regulation. Instead of strictly economic criteria, it is of course possible to use another different criterion or combination of criteria.

It should be noted that the CAMC will only act if strictly required, and will not try to globally change setpoints in order to achieve a near optimum point of operation of the whole multi-microgrid. This justifies the choice of using power setpoint variations and not absolute power setpoints. Using power setpoint variations provides a simple way to have a higher-order control system, either automatic or manual, that would independently adjust microsource or DG output to setpoints other than the system optimal ones. One example of this "control system" could be the microsource individual owners who would adjust microturbines, for instance, according to their heating needs.

5.4.1.2   Control Level 3 – MGCC

Each of the MGCCs will also allocate the necessary power changes to its subordinate controllable loads and microgeneration units, through the load controllers (LC) and micro-source controllers (MC). Some of these microsources do not usually have full regulation capabilities and will not normally be asked to change power generation. This can be the case for both PV and wind generation, due to limitations in primary resource availability. In this case, however, it is possible to consider that generation can be curtailed in case of an excess power imbalance.

5.4.1.3   Load-Shedding

The approach adopted for load-shedding, in the context of this hierarchical control system, is fairly different from the one used in conventional systems, because the hierarchical control system which supervises the controllable loads isn't capable of acting in near instantaneous time frames (mainly owing to communication system limitations). Therefore, in these circumstances, load-shedding is not expected to reduce the amplitude of the frequency excursions in the few seconds following a disturbance. It should be seen more as a kind of secondary reserve – rather than an emergency resource – helping the frequency to return to the rated value faster or without depending so much on the availability of renewable resources or other generation systems.

The main difference that distinguishes microsources from controllable loads is that it is not feasible to keep loads disconnected indefinitely (except for this detail, loads could be regarded simply as negative generation). It is thus mandatory to reconnect as much load as possible after the system returns to a near normal frequency value. This is accomplished through the use of a control loop that runs on a larger timescale, reconnecting loads after the system has been running at a near steady-state condition for a predefined period of time.

Starting from the assumption that the system – after some time of running stable and near the rated frequency – is capable of supporting the connection of further loads, the control systems starts to reconnect the most expensive/important ones first. This is done step by step, always ensuring that there is enough available reserve on the multi-microgrid in order not to unnecessarily compromise the system's stability. Before each new reconnection, the control system waits for the frequency to stabilize.

5.4.1.4   Energy Storage Systems

A cluster of several storage devices (e.g. flywheels and batteries) could, if integrated into the hierarchical control system, efficiently establish a storage reserve that would be of great help to the islanded operation of the network at both the microgrid and the multi-microgrid level. Also, we are assuming that most of these storage devices have interface inverters of the voltage source inverter (VSI) type, which can have their output power controlled on the basis of frequency droop.

Therefore, these storage devices can help in two possible ways: (a) they can act autonomously, with their output power P_{VSI} responding to system frequency changes, providing energy used to initially balance the system using a proportional control element as described next or (b) they can receive setpoints controlled from a central location, in a hierarchical way.

$$P_{VSI} = K_P \times (f_{rated} - f)$$

These two control methods are not mutually exclusive: while an autonomous response will undoubtedly improve the system's response to the initial frequency deviations following a disturbance, the hierarchical system can take over after that initial response and reallocate each source and storage element contribution according to some predefined criteria. This two-step approach can be justified by the intrinsically slow nature of the hierarchical control system, which suggests that grid-connected storage devices under hierarchical control should be regarded as a secondary reserve while, at the same time, they should be capable of acting autonomously in order to be able to limit initial frequency excursions.

5.4.2 Hierarchical Control Details

In this approach, the system's frequency is continuously monitored by the CAMC (Figure 5.11). Every time interval Ts (sample time), if triggered by significant changes in frequency, the CAMC will send control setpoints to every MGCC, other DG units and controllable loads. This sample time Ts cannot be very small, mainly because of the constraints imposed by the communication system on which this control system depends.

Therefore, the frequency error and the frequency error integral will be used to determine the additional power ΔP to be requested to the available contributors under CAMC control: MGCCs, DG units and controllable loads. This approach effectively implements a kind of PI controller.

$$\Delta P = \left(K_P + K_I \frac{1}{s} \right) \times (f_{rated} - f) \tag{5.8}$$

It should be noted that this additional power can have negative values if the frequency rises above its rated value. In this way, the CAMC can also respond to other disturbances, such as load loss while in islanded mode, instructing the distributed generation to reduce power output (including microgeneration curtailment, if necessary), eventually reconnecting some loads still disconnected at the moment.

If the required power variation ΔP is larger than a predefined threshold (related to a deadband), the control system will proceed to determine how to optimally distribute the power requests through the available sources. Unitary generation costs for each of the sources (MGCCs and other DG units) can be used for this purpose.

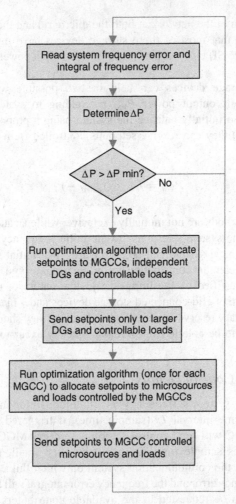

Read system frequency error and
integral of frequency error

Determine ΔP

$\Delta P > \Delta P$ min?

No

Yes

Run optimization algorithm to allocate
setpoints to MGCCs, independent
DGs and controllable loads

Send setpoints only to larger
DGs and controllable loads

Run optimization algorithm (once for each
MGCC) to allocate setpoints to microsources
and loads controlled by the MGCCs

Send setpoints to MGCC controlled
microsources and loads

Figure 5.11 Implementation flowchart

The optimization can be based on standard linear optimization techniques:

$$\min_x z = c^T x \qquad (5.9)$$
$$\text{subject to} \quad \sum x = \Delta P \qquad (5.10)$$
$$x \geq b_1 \qquad (5.11)$$
$$x \leq b_2 \qquad (5.12)$$

Where the vectors represent

c generation cost and load curtailment prices
x generation or load setpoint variations
b_1 smallest variations allowed (lower bounds)
b_2 largest variations allowed (upper bounds).

The set of restrictions can also define which generators/loads participate in frequency regulation. This can be done by setting to zero the ith elements of both b_1 and b_2, corresponding to units that cannot be adjusted.

Because loads are considered as negative generation, the corresponding coefficients (elements in vector c of prices) are negative.

In order to avoid globally changing setpoints (e.g. decreasing production from expensive microsources and replacing them with less expensive ones), it is necessary to adjust the lower and upper bounds in the optimization procedure according to the ΔP value:

$$\begin{cases} \Delta P > 0 \Rightarrow b_1 = 0 \\ \Delta P < 0 \Rightarrow b_2 = 0 \end{cases} \tag{5.13}$$

The enforcement of the above described conditions ensures that no microsource will decrease its production, so that another one can increase it (i.e. there will not be any unsolicited power transfers between microsources).

This optimization is performed each sample period Ts and will originate a vector representing the power generation changes to be requested from microgrids (MGCCs), independent DG units (e.g. combined heat and power – CHP) and loads (MV load-shedding operations).

Each MGCC will now use the power change requested by the CAMC to establish the main restriction of a new optimization procedure, which could be identical to the one used before by the CAMC. This optimization will now help determine the power changes to be requested from microsources and controllable loads under MGCC control, thus one level below in the hierarchical control structure.

5.4.3 Main Results

Using the described coordinated frequency control methods, the system is able to handle severe network disturbances such as islanding events. It can also perform load-following by scheduling power generation among generation units and microgrids under its control. A dynamic simulation platform was developed for evaluation purposes and includes an extended version of the multi-microgrid test network shown in Figure 5.12 [16].

Figure 5.13 illustrates the frequency control approach described, by showing the response of the hierarchical control system to a sudden islanding and also to gradual load changes in islanded mode. In this test case, the MV multi-microgrid network has a total load of 19.9 MW and is importing approximately 5.3 MW of active power from the upstream HV network.

The test disturbances include the disconnection of the HV/MV branch, effectively islanding the multi-microgrid system, and also a load change at several nodes at a rate of 4% per second for 10 s (for total increase of nearly 0.9 MW). These disturbances are also pictured in Figure 5.13.

The results show how the hierarchical control adopted in this multi-microgrid manages to recover the frequency to the rated value after islanding. Although the minimum frequency value after the disturbance remains practically unaltered, in the moments that follow, the system evolves in a much better way.

The success of the frequency recovery is in part due to the fact that the CAMC is sending setpoints to the microgrids and other DG units dispersed on the MV network. The

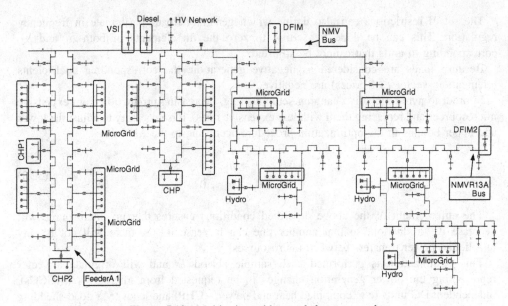

Figure 5.12 Multi-microgrid extended test network

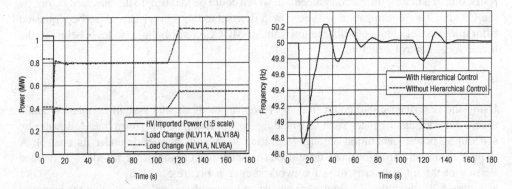

Figure 5.13 Test disturbances and frequency behavior with and without secondary control

microsources inside the microgrids are also subject to setpoint attribution according to a similar price-dependent optimization algorithm. Figure 5.14 shows the microgrid setpoint modifications and how they are directly translated into individual setpoints for each of the microsources within that same microgrid.

When the multi-microgrid system does not have sufficient reserves to respond to a large loss of imported power, load curtailment may be required. This control system is designed to support these actions as well, as demonstrated in Figure 5.15.

The hierarchical control algorithm is thus capable of acting on controllable loads, integrating them into the optimization process. The expected benefits include the increase in system response speed (Figure 5.15), while still complying with the optimization rules in use.

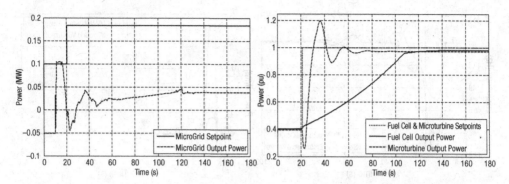

Figure 5.14 Example microgrid external and internal setpoint commands and output power

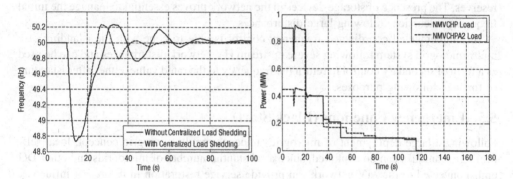

Figure 5.15 Influence of load shedding and shedding steps during islanding

Loads are disconnected according to their "cost" or importance. Figure 5.15 (right) illustrates how load-curtailment occurs, showing two of the larger loads, but considered less important. These loads begin to be curtailed soon after the main disturbance (islanding). On the other hand, more expensive/important loads would begin to be disconnected later, after all the less expensive ones have attained their minimum values.

As mentioned, the test network contains several VSI devices coupled to storage elements. The output of VSI is based on the frequency error through the use of a proportional controller. Figure 5.16 illustrates VSI influence on the frequency variation of the sample test system, together with an example of the power output of a large VSI (connected at the MV level) and one of the microgrids' VSIs. All the previously shown simulation results benefit from the usage of the storage elements as active and autonomous participants in frequency regulation. It should be noted that the VSIs are also programmed to emulate the behavior of unregulated synchronous machines, which may mask the influence of the proportional controllers during the initial phase of heavier disturbances.

These examples show how the setpoint modification commands sent to DG units, micro-grids and controllable loads enable the frequency to return to the rated value in a reasonable amount of time. In addition, the control system can resort to load-shedding, in order to better support the transition to islanded operation in scenarios of low multi-microgrid power

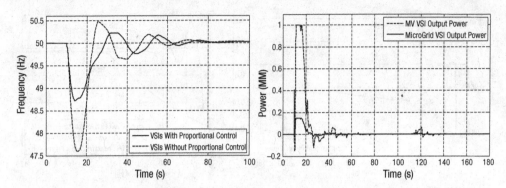

Figure 5.16 Influence of autonomous proportional control in storage devices

reserves. The presence of storage devices in the network proves essential to manage the initial frequency excursion following large disturbances.

The implemented coordinated frequency control is able to cope with the islanding of a multi-microgrid system following severe upstream HV network disturbances and can be used as a form of secondary control to return the frequency to the rated value, either after islanding or for load-following purposes.

5.5   Emergency Functions (Black Start)

Following a large deployment of microgrids under the multi-microgrid concept, local self-healing techniques can be exploited, since a substantial number of microgrids and other DG units connected to the MV network can provide service restoration in its area of influence. Therefore, after a local blackout or a general blackout where the multi-microgrid was not able to isolate and continue to operate in islanded mode, it is possible to achieve a reduction of interruption times by exploiting multi-microgrid black start capabilities. These capabilities provide a fast black start concurrently at both LV and MV levels and allow multi-microgrid system islanded operation until the upstream HV network is available.

In fact, concerning the microgrid concept [17] the interruption time of LV consumers can be reduced by allowing microgrid islanded operation until the MV network is available, thus exploiting microgeneration control capabilities to provide fast service restoration at the LV level. This first step will be followed by the microgrid synchronization with MV grid when it is available. Based on microgrid control strategies and making use of the microgrid communication infrastructures, special issues for microgrid service restoration were identified in order to totally automate the microgrid restoration procedure [18,19]. Therefore, during multi-microgrid service restoration it was assumed that the MGCC of each microgrid is responsible for its service restoration and will build up the microgrid system autonomously.

Because a significant part of the full multi-microgrid restoration process can be performed in several islands in parallel, the time between grid blackout and power restoration to a substantial proportion of the consumers can be much reduced. However, the final goal of restoration is to reconstruct the multi-microgrid state existing before the loss of power, and this usually includes avoiding isolated islands, even if they are semi-autonomous microgrids. Therefore, the CAMC must also play an important role in coordinating the interconnection of all the temporary autonomous islands.

In a multi-microgrid, the rules required to accomplish a successful restoration can be more general than those that can be adopted in a HV/VHV network. This fact is almost unavoidable because MV networks can be extremely diverse as regards network characteristics and types of energy sources, power electronics interfaces and loads. Therefore, it can become difficult to find a very specific set of rules that fits every network and every operation point of a network.

The black start software module to be implemented on the CAMC is responsible for managing this set of rules and conditions to be checked during restoration. These rules and conditions define a sequence of control actions to be carried out during multi-microgrid restoration and must be coordinated and planned beforehand. For this purpose, a sequence of control actions can be identified and evaluated through numerical simulation. The proposed sequence of control actions, to be stored at the CAMC level, allows the definition of a procedure that aims at the full automation of the entire multi-microgrid service restoration procedure and subsequent synchronization with the upstream HV system.

5.5.1 Restoration Guidelines

Multi-microgrid restoration will be triggered if a general or local blackout occurs or if major disturbances affecting the HV upstream system do not allow the multi-microgrid to be supplied from the HV side after a predefined time interval. The CAMC should then be able to receive information from the DMS about the service restoration status at the HV level in order to decide about launching a black start procedure. In addition, the CAMC will guide multi-microgrid service restoration, based on information about the last multi-microgrid load and generation scenarios. This information is kept in a maintained database together with the information about restart availability of DG units connected to the MV network and is used to further adjust and refine the sequence of control actions, mostly identified in advance and embedded into the CAMC software, which will finally lead the system to a state similar to the pre-fault scenario.

The CAMC is also used for secondary frequency control during load-following conditions when the multi-microgrid is operated in islanded mode, as presented in Section 5.4. In order to balance generation and load, the CAMC generates setpoints based on system frequency changes and sends them to the MGCC and to controllable DG units. In turn, the MGCCs will act as intermediate controllers between the CAMC and the microgrid internal controllers. MGCCs are responsible for scheduling the corresponding power change among the micro-generation units by sending setpoints to the corresponding controllers.

The MGCC is also responsible for guiding the microgrid black start procedure, and so it is assumed that microgrid service restoration is carried out autonomously when the microgrid is operated in islanded mode until the MV network is available. As several islands will be formed during the early stages of the multi-microgrid black start procedure, it is also assumed that the automatic secondary frequency control system embedded into the CAMC software is initially turned off. This allows a simpler coordination of the restoration procedure, while ensuring that no contradictory commands ever arise.

Thus, the information exchange between the CAMC and local controllers during the multi-microgrid black start procedure will mainly involve switching orders. The purpose of these commands is to rebuild the MV network and also to connect DG units, microgrids and MV loads, according to general monitoring information (e.g. voltage levels, system frequency and grid status). All this information is received and processed by the CAMC through the communication infrastructure that is supposed to be available. Verification of synchronization conditions is performed locally.

Another basic requirement is the availability of DG units connected to the MV network with black start capability. Its restart procedure is carried out prior to network energization, so it is not reflected in the MV network. The same happens with the several microgrids with black start capability.

Furthermore, it is also assumed that it is possible to prepare the network for energization. To accomplish this, after system collapse, the following requirements should be taken into account:

- the multi-microgrid is disconnected from the upstream HV network,
- multi-microgrid feeders are fully sectionalized,
- microgrids are disconnected from the MV network,
- DG units and loads are disconnected from the MV network,
- the HV/MV transformer is disconnected from the HV and MV networks,
- all MV/LV transformers are disconnected from the MV and LV networks,
- all the reactive power sources, such as shunt capacitor banks, are switched off.

From this starting point, the multi-microgrid restoration procedure is carried out with the aim of supplying consumers as soon as possible, while satisfying the system operation conditions. So, after a general blackout, the CAMC will perform service restoration in a multi-microgrid based on information stored in a database about the last multi-microgrid load scenario, as described before, by performing the following generic sequence of actions (the first two steps include the actions described in the previous paragraph):

1. *disconnecting all loads, sectionalize the corresponding MV/LV transformers and switching off the reactive power sources* – after a general blackout all the loads, transformers and shunt capacitor banks should be disconnected in order to avoid large frequency and voltage deviations when energizing the MV network.
2. *sectionalizing the multi-microgrid around each microgrid and around each DG unit with black start capability* – this leads to the creation of small islands inside the multi-microgrid, since after their autonomous black start, microgrids keep operating in islanded mode feeding some amount or its own entire load. Other higher powered generators may be also supplying their protected loads (e.g. some diesel groups and CHP units). These islands will all be later synchronized together.
3. *building the MV network* – a diesel group or other black start capable generator (most probably a synchronous machine) is used to energize the initial part of the MV network, which comprises some unloaded transformers downstream and some paths that allow the synchronization of other units feeding their own loads (e.g. CHP islands) or feeding important loads. The energization of the initial part of the MV network is carried out step by step in order to avoid large voltage and frequency deviations.
4. *synchronization of islands with the MV network* – each one of the existing islands can now be synchronized with the MV network when the corresponding path is energized, in order to strengthen the multi-microgrid system. The synchronization conditions (phase sequence, frequency and voltage difference) should be verified in order to avoid large

transient currents. As most of the microgrids are capable of autonomous black start – even if in limited operating conditions – they may not need to be connected at this point.

5. *connection of a certain amount of important load* – connection of important loads is performed if the DG units connected to the MV network are capable of supplying these loads. The amount of power to be connected should take into account the generation capacity in order to avoid large frequency and voltage deviations during load connection.

6. *energization of the remaining MV branches and the MV/LV transformers upstream of the microgrid* – at this stage the islands containing isolated DG units are already synchronized and the multi-microgrid is strong enough to energize the remaining branches of the MV network. However, as the multi-microgrid comprises a large number of MV/LV unloaded transformers, their energization should be carried out in several steps in order to avoid large inrush currents. Thus, the MV/LV transformers upstream of the several microgrids are energized first, in order to allow for the microgrid synchronization with the MV network.

7. *synchronization of microgrids with the MV network* – microgrids operated in islanded mode can then be synchronized with the MV network. For this purpose the synchronization conditions should be verified.

8. *energization of the remaining MV/LV transformers* – in order to be possible to completely restore the load, all the MV/LV unloaded transformers should be energized. They are divided into several groups which are energized at different times in order to avoid large inrush currents. Afterwards the MV/LV transformers upstream of the uncontrollable DG units are also energized.

9. *load restoration* – at this stage the MV network is fully energized and some loads can be connected depending on the generation capacity.

10. *connection of uncontrollable DG units connected to the MV network* – at this stage it is supposed that multi-microgrid becomes sufficiently strong to smooth voltage and frequency variations due to power fluctuations in non-controllable DG units, allowing their connection to the MV network. MV paths are also created, so that DG units without black start capability can absorb power from the grid in order to restart.

11. *load increase* – in order to feed as much load as possible, other loads can then be connected.

12. *activation of the automatic frequency control* – the automatic frequency control is now activated in order to ensure that the multi-microgrid system frequency is near its nominal value while the multi-microgrid is operated in islanded mode.

13. *multi-microgrid reconnection to the upstream HV network when it becomes available* – the synchronization conditions should be verified again, after the synchronization order is given by the CAMC. The HV/MV transformer should be previously energized from the HV side and the synchronization is performed through MV switches.

The technical feasibility of the proposed sequence of control actions, which allows multi-microgrid service restoration was evaluated using the test network presented in Figure 5.12.

5.5.2 Sample Restoration Procedure

The feasibility of the proposed sequence of actions to carry out multi-microgrid service restoration is demonstrated in this section through numerical simulations.

According to the proposed sequence of actions, the multi-microgrid system restoration procedure can be split into the next two main parts:

- MV network energization and synchronization of small islands and
- load supply and integration of generation.

In the first part, the skeleton paths of the MV network are energized and DG units supplying their protected loads can be synchronized. Then, some load should be restored in order to balance the generation and to stabilize the voltage. At this stage, the main problems to deal with are mainly the voltage profile and switching operations as a consequence of energizing unloaded MV paths and a large number of unloaded transformers.

In the second part, load is restored according to generation requirements, and microgrids can be synchronized. Other DG units can also be connected to the MV network. Then, the main problems to deal with concern active and reactive power balance, overloads and the response of prime movers to sudden load pick-up.

Considering the studied test network, the diesel and CHP units can restart successfully without network support. It can even be assumed that these units are already running and feeding their own loads in the first stages of the black start procedure. The diesel group was selected to energize the initial part of the MV network in order to create paths to connect the CHP units. Later, the islands formed by both CHP1 and CHP2 units are synchronized with the MV network and the MV/LV distribution transformers of Feeder A1 are energized. In order to prevent large inrush currents and therefore voltage drops inside the energized multi-microgrid system, this task should be performed in stages. Thus, three groups of transformers were considered and energized at different times. The impact of these control actions on the MV network can be observed in Figures 5.17 and 5.18. The adopted procedure allows the system frequency and bus voltages to be kept within acceptable limits.

When the multi-microgrid network is fully energized, and all CHP units are synchronized with the MV network, the service restoration proceeds with load supply and integration of other DG units. Load is restored step by step according to the integrated generation capacity. As the secondary frequency control is disabled, every load pick-up results in a system frequency drop. Therefore, it is necessary to connect further DG units as the amount of

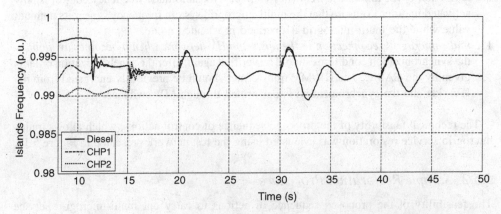

Figure 5.17 Frequency following islands synchronization and feeder A1 energization

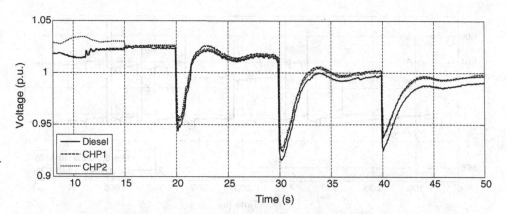

Figure 5.18 Bus voltages following islands synchronization and feeder A1 energization

restored load increases. For this purpose, when the CAMC observes system frequency higher than its nominal value, power is available to connect additional load and therefore, depending on the system capacity, the CAMC determines the amount of additional load to be connected. After supplying all the loads, the secondary frequency control is activated and the system nominal frequency is restored. The feasibility of these control actions is illustrated in Figures 5.19 and 5.20.

The results obtained demonstrate the technical feasibility of the multi-microgrid concept improving service restoration procedures on distribution systems allowing the reduction of load restoration times. Islanded operation of parts of the MV grid during the restoration sequence has a key role in the success of this procedure. In addition, multi-microgrid service restoration can be considered as a great opportunity to take advantage of generation systems, traditionally considered as non-controllable, as support to the load supply. Based on the multi-microgrid hierarchical control system and exploiting its communication infrastructure, the entire multi-microgrid service restoration procedure can be fully automated.

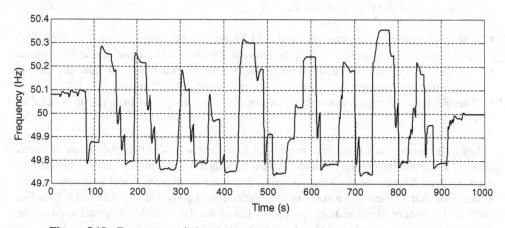

Figure 5.19 Frequency variation during load supply and generation increasing stage

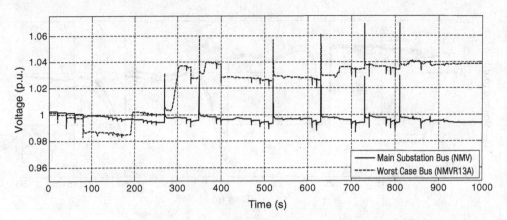

Figure 5.20   Voltage variation during load supply and generation increasing stage

5.6 Dynamic Equivalents

Large deployment of microgrids within the framework of the multi-microgrid concept presented previously will lead to very large MV systems comprising hundreds or even thousands of these active cells formed at the LV levels and connected to the MV network, together with other DG units. In order to operate the microgrid within the expected reliability levels, focusing specially on the multi-microgrid autonomous mode of operation, transient and dynamic stability studies need to be performed at the MV level. Thus, multi-microgrid islanding and load-following are key studies to be performed within this framework. However, using detailed mathematical models able to accurately represent the microgrid dynamic behavior with impact on the MV level means dealing with very high dimension systems, with a computational burden that will render multi-microgrid dynamic behavior studies unfeasible. Therefore, deriving dynamic equivalents for microgrids is required to speed up this type of numerical simulation.

The microgrid reduced-order models are intended to replace the microgrid detailed models according to the following guidelines:

- The microgrid dynamic equivalent should be an accurate representation of the corresponding detailed model, regarding the relevant dynamics with impact on the MV system.
- The cost of building the dynamic equivalent must be much smaller than the cost of performing the transient analysis using the microgrid detailed model.
- The obtained microgrid dynamic equivalents will be integrated in dynamic simulation tools.

Since the microgrid's main features do not lend themselves to the application of modal analysis and coherency based methods, system identification techniques can be exploited for deriving dynamic equivalents for microgrids. Since the equivalent model is developed in a system-oriented framework, a solid system definition is provided in Section 5.6.2, before starting the system identification procedure. Based on the available physical knowledge, effectively used, two suitable approaches are formulated, relying on using either a black box model structure or a physical model structure. The first tries to exploit the full response of the microgrid, while the second tries to understand the physical behavior of the different components of the microgrid.

5.6.1 Application of Dynamic Equivalence Based Approaches to Microgrids

Typical approaches to deriving dynamic equivalents for power systems rely basically on system reduction and system identification based techniques [20,21]. Reduction techniques are based on aggregation and elimination of some components of the system detailed model by exploiting modal analysis [22,23] and coherency based aggregation methods [24–26]. In the system identification based approaches, the dynamic equivalents are derived from data, using a dataset comprising either simulated data or measured data collected at specific points of the system. The parameters of the model are then adjusted, so that the model response matches the observed data. Artificial Neural Networks (ANNs) have been the most prevalent system identification based method, because of their high inherent ability in modeling nonlinear dynamic systems, being the dynamic properties obtained only from data [27–31].

When compared with conventional power systems, microgrid systems have no centralized synchronous machines. Rather, despite their lower physical dimensions, the microgrid comprises a number of small-scale DG units, with different technologies, connected to the LV network through power electronic interfaces and exhibiting nonlinear dynamic behavior. Also, different microgrids will have quite different compositions and obtaining detailed information about all of their components will be a very difficult task. Thus, the application of system reduction based approaches for deriving dynamic equivalents for microgrids have the following main drawbacks:

- Modal analysis requires performing very time-consuming procedures and the obtained reduced-order models lack accuracy when the system steady-state operating conditions move away from the base case.
- Coherency based methods are suitable only for conventional power systems with synchronous generators concentrated into a few areas. Moreover, the term coherency becomes less meaningful since power electronic based interfaces can almost completely separate the dynamic behavior of generators from the network, resulting in quite different dynamics of DG units, strongly influenced by the control systems of power electronic interfaces. Also, some DG technologies do not have rotating parts – for example, fuel cells and PV systems.

Due to its general applicability, system identification based approaches have been exploited for deriving aggregated models for distribution networks with large-scale integration of DG [30,31]. The ANN based approaches have been the key players in this endeavor, since they can learn nonlinear maps from data. This feature, together with the introduction of powerful optimization tools, allows the handling of a wide range of nonlinear dynamics, including the relevant dynamics of microgrids [32], lacking detailed physical knowledge. This represents a significant advantage, especially when there is a limited understanding of the relations between system variables.

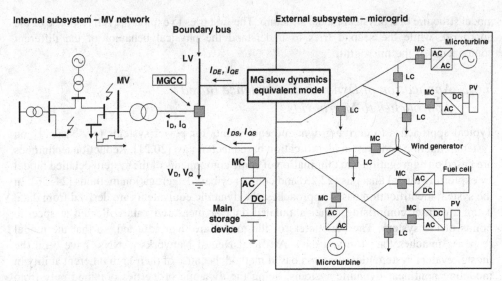

Figure 5.21 multi-microgrid equivalent model including the microgrid dynamic equivalent

5.6.2 The Microgrid System Definition

For the purposes of analysis, and to get a better perspective for setting up a reduced equivalent system, the detailed model of the whole multi-microgrid system is divided into two main parts [32–34]: (1) the internal area to be retained for detailed analysis and the external area to be replaced by the equivalent model, as depicted in Figure 5.21, (2) the dynamic system to be identified, consisting of a set of differential and algebraic equations describing the state evolution over time of the physical system – the microgrid.

Detailed models of microgrids focus on the DG technologies, on their power electronic based interfaces and on the control strategies suitable for operating the inverter dominated microgrid. Power electronic converters are represented by their control functions only, implementing the VSI and PQ control, so that switching transients, harmonics and inverter losses are neglected [3,18,35–38]. Also, three-phase models have been used for both DG units and power electronic based interfaces, only considering microgrid three-phase balanced operation. It was also assumed that the microgrid is operating within the framework of a single master operation (SMO) control strategy [3]. By performing time domain simulations using the microgrid detailed models, two different time scales can be distinguished:

- VSI interfacing the microgrid main storage device, with fast dynamic responses and
- small-scale DG systems interfaced through PQ inverter controls, with slow dynamic responses.

Therefore, suitable dynamic equivalents for microgrids comprise two main parts:

- the microgrid main storage device, represented as a constant DC voltage source behind the VSI detailed model and
- the microgrid slow dynamics equivalent model representing the aggregation of all the remaining components.

Then the aim of the system identification procedure is to identify the equivalent model able to represent the microgrid slow dynamic behavior, as represented in Figure 5.21. Thus, based on the system definition and using the engineering expertise, the numerical set-up was implemented – a dedicated dynamic simulation platform able to design suitable numerical experiments with microgrids, in order to produce sufficiently informative datasets. Also, the simulation platform should be able to evaluate the performance of the microgrid reduced-order models at the final stage of the system identification procedure.

Then, the microgrid detailed model is excited through simulated disturbance scenarios into the MV network, such as multi-microgrid islanding and load-following when the multi-microgrid is operated autonomously, being the microgrid dynamics captured by means of both the input and output electrical signals measured at the system boundary according to a suitable sample time. Boundary bus voltages – expressed in the D-Q synchronous reference frame – and system angular frequency are considered as inputs, while the boundary bus injected currents in the tie lines, also expressed in the D-Q reference frame, are considered as outputs. Node elimination and aggregation is performed by making the injected currents of the aggregated model equal to the currents in the tie-lines. Thus, the microgrid slow dynamics equivalent model is disturbed by both boundary bus voltage and system frequency variations, reacting by varying the injected currents into the boundary bus, operating according to the principles of a Norton model [34].

5.6.3 Developing Micogrid Dynamic Equivalents

Developing reduced-order models, using system identification based procedures, requires the following main tasks to be performed:

- data generation,
- model structure selection,
- identification method selection,
- model validation.

However, nonlinear system identification based techniques are application dependent. General guidelines recommend using the available prior knowledge as well as the engineering expertise to derive models as close as possible to their intended purpose. Regarding the model structure selection, the physical insights about the microgrid dynamic behavior have been combined with the formal properties of the models, yielding two promising approaches:

- black-box modeling based on a time delay neural network (TDNN) model structure, aiming to exploit the full response of the microgrid when excited after a disturbance and
- physical modeling, seeking to understand the physical behavior of the different components of the microgrid.

5.6.3.1 TDNN Based Dynamic Equivalents

The TDNN based on the multi-layer perceptron (MLP) neural networks have high capability for dealing with complicated nonlinear problems. When well-trained, the TDNN can replace the microgrid's slow dynamics and it is expected that it will properly interact with the retained network for a wide range of operating conditions. In the following, we describe the main tasks to be performed within the framework of system identification based procedures using TDNN to develop microgrid dynamic equivalents.

Data Generation

The dataset is almost the only source of information for building the TDNN based dynamic equivalent model. Thus, a detailed model of the multi-microgrid is used for generating a sufficiently informative dataset, comprising a large number of samples, in order to form appropriate training and validation datasets. Thus, boundary bus voltages, system frequency and injected currents are stored in a database during the simulation of the disturbances, in order to build suitable training patterns. Since signals are likely to be measured in different physical units, it is recommended to remove the mean and scale all signals to the same variance, in order to avoid the tendency for the signal of largest magnitude to be too dominating. Moreover, scaling makes the training algorithm numerically robust and leads to a faster convergence, and also tends to give better models [39]. In order to generate a more robust TDNN, which is able to simulate the microgrid dynamic behavior under different operating conditions, normalized deviations of voltage, system frequency and currents from the corresponding steady state are used, similarly to the approach proposed in [30,31].

Thus, a function f_1 computes the normalized voltage (Δv_D, Δv_Q) and system frequency deviations ($\Delta \omega$) while a function f_2 computes the current to be injected into the retained network (I_{DR}, I_{QR}), as follows:

$$f_1 : \Delta v_D = \frac{V_D - V_D^{(0)}}{\Delta V_{D,\max}} ; \Delta v_Q = \frac{V_Q - V_Q^{(0)}}{\Delta V_{Q,\max}} ; \Delta \omega = \frac{\omega - \omega^{(0)}}{\Delta \omega_{\max}} \qquad (5.14)$$

$$f_2 : I_{DR} = \Delta i_D \times \Delta I_{D,\max} + I_{DR}^{(0)} ; I_{QR} = \Delta i_Q \times \Delta I_{Q,\max} + I_{QR}^{(0)} \qquad (5.15)$$

Where $\Delta V_{D,\max}$, $\Delta V_{Q,\max}$ are the maximum deviations of the direct and quadrature components of the voltage, $\Delta I_{D,\max}$, $\Delta I_{Q,\max}$ are the maximum deviations of the direct and quadrature components of the injected current and $\Delta \omega_{\max}$ is the maximum deviation of the system frequency, regarding its nominal value, ω_0. In turn, $V_D^{(0)}$, $V_Q^{(0)}$ represent the steady-state voltage in the direct and quadrature components, and $I_{DR}^{(0)}$, $I_{QR}^{(0)}$ represent the steady-state current in direct and quadrature components.

The initial steady-state values of boundary bus voltage and injected current of the microgrid slow dynamics equivalent model are determined through the initial load flow calculations. Their maximum deviations, as well as the maximum frequency deviation, are obtained from the dynamic simulation of the largest amount of load connection and disconnection upon multi-microgrid islanding.

Model Structure Selection

TDNN represents the microgrid slow dynamics according to the external dynamics based approaches [39]. Then, the nonlinear model comprises the regression vector and nonlinear mapping that combines the regressors into a one-step-ahead prediction, as represented in Figure 5.22.

Therefore, selecting the TDNN model structure requires selecting a particular structure of the regression vector and subsequently to specify the number of hidden units, in an

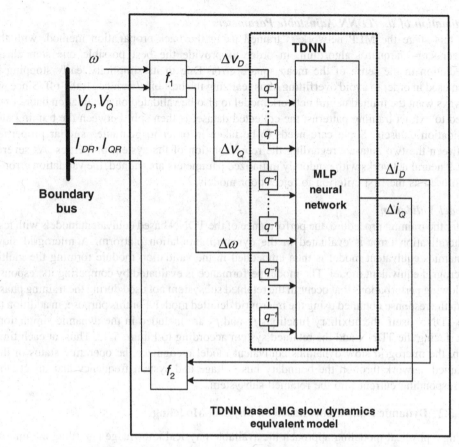

Figure 5.22 TDNN based microgrid slow dynamic equivalent model

attempt to determine good network architectures for this choice of regressors. Taking into account the intended purpose of the microgrid dynamic equivalent, it should be noted that past system outputs are not available. Thus, the adopted structure of the regression vector is based on the nonlinear finite impulse response (NFIR) model, considering the following assumptions:

- The nonlinear dynamic model output was not considered, to avoid problems of error accumulation and error instability in time domain simulations.
- The model order results from a trade-off solution between the input space dimensionality and the number of time delays that allow representing the dynamics of interest.

The definition of the MLP structure comprises only the definition of the hidden units and the choice of the activation functions, being the number of regressors as well as the number of hidden units obtained further, within the framework of training and validation stages, by using trial and error approaches.

Estimation of the TDNN Adjustable Parameters

At this stage the MLP network is trained using the back-propagation method, with the Levenberg–Marquardt algorithm, in order to provide the best possible one step ahead prediction in the sense of the mean square error. Due to its simplicity, early stopping is also used in order to avoid overfitting, thus realizing the best bias/variance trade-off. Since we always want the trained neural network model to also be validated on a validation dataset not used to extract training patterns, the collected dataset is then split between the training and validation datasets. Some care needs to be taken in order to guarantee similar properties between the two datasets regarding the representation of the system properties. As several MLP neural networks with randomly initialized parameters are trained, the validation error is also used as the first criterion to reject poor models.

Model Validation

After the training procedure, the performance of the TDNN based equivalent models with less generalization error is evaluated in the dynamic simulation platform. A microgrid slow dynamics equivalent model is then embedded in the validation module forming the multi-microgrid equivalent model. The model performance is evaluated by comparing its response following perturbations that occur in the retained subsystem not used during the training phase with the response obtained using the microgrid detailed model. For this purpose, in addition to the TDNN itself, the auxiliary functions f_1 and f_2 are included in the dynamic simulation, interfacing the TDNN and the retained system according to Figure 5.22. Thus, at each time step, the microgrid slow dynamics equivalent model recognizes the operating status of the retained network through the boundary bus voltage and system frequency and injects the corresponding current into the retained subsystem.

5.6.3.2 Dynamic Equivalents Based on Physical Modeling

Using a physical modeling approach the available physical knowledge regarding the microgrid composition and microgrid slow dynamic behavior is explicitly incorporated within the model structure and, therefore, the requirements to build the dataset are less demanding. Thus, the system identification main stages are related to the model structure selection and to the identification method, as described next.

Model Structure Selection

The physical laws that approximate the microgrid slow dynamics under study are similar to those that govern the active power control in a diesel engine [33], reacting to the system frequency variations by changing the output power, as can be seen in Figure 5.23.

Since the microgrid slow dynamics reduced-order model is required to behave according to the principles of the Norton model, the instantaneous power theory [40] was used in order to determine the network injected current. For this purpose, only physical laws are used without parameterization, so that the parameters of the physical model structure whose values have to be estimated during the identification procedure are gathered into the parameter vector, θ, as

$$\theta = [R \quad K_1 \quad K_2 \quad T_2 \quad T_D] \tag{5.16}$$

The block of instantaneous power theory requires only the implementation of the algebraic equations needed for computing the current injected on the retained subsystem

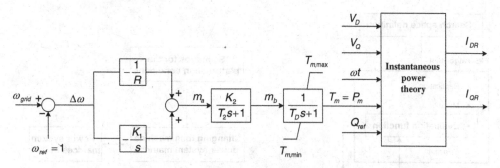

Figure 5.23   Model structure of the microgrid slow dynamics equivalent model

(I_{DR}, I_{QR}), based on the boundary bus voltage (V_D, V_Q) and on the active power delivered by the microgrid slow dynamics equivalent model, P_m. In addition, a given reactive power is used, Q_{ref}, corresponding either to a predefined value linked to the power factor of the DG units or to a reactive power setpoint sent by the MGCC, according to the SMO control strategy [3].

The Identification Method

In order to estimate the parameters of the physically parameterized model structure, a suitable identification method is required. For this purpose the EPSO tool [11,12] was adopted as a global optimization tool together with the sum square error (SSE) criterion [33,34]. In this context, the vector of parameters (θ) provides the particle phenotype descriptions corresponding to the particle positions into the predefined parameter space. The parameter estimation procedure is performed online. For each particle, the loss function expressed in terms of the SSE is evaluated by performing time domain simulations. Therefore, some interaction between the EPSO algorithm and the multi-microgrid equivalent model is required, as can be seen in Figure 5.24.

After defining the search space through both the minimum and maximum values of each parameter into the parameter vector, it is expected that the EPSO algorithm will perform the search to the global optimum or, at least, to a good local optimum in the SSE sense. For this purpose, after mutation has been performed by the EPSO algorithm, the following sequence of steps has to be carried out for each particle in the swarm:

1. The evaluation function sends the particle object parameters to the microgrid slow dynamics equivalent model.
2. A pre-specified set of disturbances occurring at defined time instants is simulated over a certain time period.
3. The microgrid equivalent model response is compared with the target response, which was generated from the microgrid detailed model, yielding an error sequence.
4. The SSE is then calculated and sent back to the evaluation function.

Based on the SSE magnitude, the EPSO algorithm performs selection in order to build the swarm corresponding to the next generation. The above procedure is repeated, while the EPSO algorithm termination condition is not verified. Then, it is expected that the microgrid slow dynamics equivalent model thus obtained will present the best performance in time

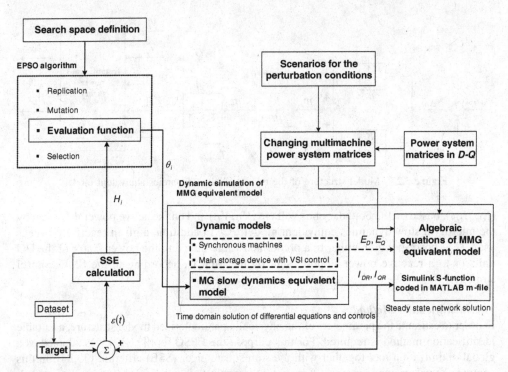

Figure 5.24 Flowchart of the parameters estimation procedure in microgrid physical modeling

domain simulations, so that the validation procedure is embedded into the parameter estimation stage.

It should be mentioned that, for a fixed and known value of Q_{ref}, fitting the current injected into the retained subsystem is similar to fitting the injected active power. Therefore, the dataset is created by the microgrid slow dynamics subsystem active power and the SSE is defined in this sense.

5.6.4 Main Results

The approaches presented in the previous sections are used to derive dynamic equivalents for microgrids using suitable test systems representing the multi-microgrid system. The microgrid slow dynamics equivalent models are connected to the system boundary bus in parallel with the microgrid main storage device, in order to replace the microgrid detailed model, thus forming the multi-microgrid equivalent model. The performance of the microgrid dynamic equivalents is evaluated by comparing the time domain responses of both the multi-microgrid detailed model and the multi-microgrid equivalent model, considering disturbances not used for extracting training patterns. Also, new initial steady-state operating conditions were considered, regarding the load and generation levels inside the microgrid and at the MV distribution network. Both microgrid dynamic equivalent models present a very high accuracy, with computational time savings, demonstrating the feasibility of the pursued system identification based approaches.

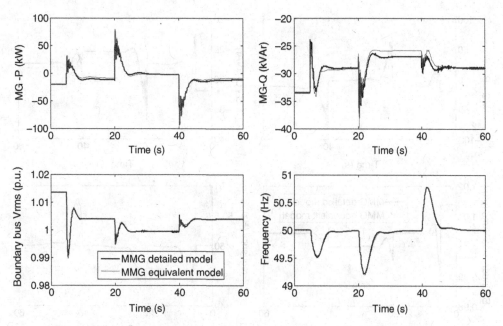

Figure 5.25 Performance of the TDNN based microgrid dynamic equivalent: behavior of microgrid active and reactive power, boundary bus voltage and system frequency

Regarding the TDNN based microgrid dynamic equivalent, the worst-case scenario was verified when the initial steady-state operating conditions moved from the base case by modifying the load inside the microgrid. For this scenario the sequence of actions was simulated as:

- multi-microgrid islanding at $t = 5$ s,
- connection and disconnection of an amount of load at $t = 20$ s and at $t = 40$ s, respectively.

The comparison between the TDNN based microgrid dynamic equivalent and the microgrid detailed model power responses can be seen in Figure 5.25. It can be concluded that this reduced-order model is effective in representing the dynamic behavior of the microgrid detailed model. Although the reactive power behavior presents a small loss of accuracy, this effect is quite small on the retained subsystem as can be observed from the boundary bus voltage and the system frequency behavior. The use of normalized deviations extends the TDNN capability of generalization to represent the microgrid dynamic behavior when the steady-state operating conditions move away from the base case. However, a very large computational effort is required to derive this reduced-order model and its domain of validity is restricted to the microgrid composition used for generating the dataset. Replacing the microgrid requires a new training procedure. These weaknesses have been overcome by using the microgrid physical model.

Then, a physical microgrid dynamic equivalent was derived using another test system comprising a microgrid with quite different composition and new steady-state operating conditions. The obtained reduced-order model of the microgrid was used to replace the microgrid in the same test system used in the case of TDNN based microgrid dynamic

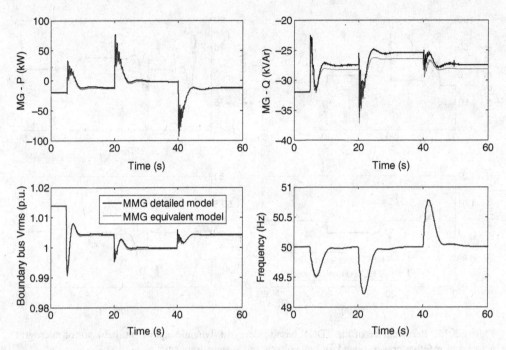

Figure 5.26 Performance of the physical microgrid dynamic equivalent: behavior of microgrid active and reactive power, boundary bus voltage and system frequency

equivalents. In order to compare the performance of the two models, the same scenario of operation was used and the results obtained are presented in Figure 5.26, which shows that the models have similar performance.

The response of the physical microgrid dynamic equivalent is in good agreement with that obtained using the multi-microgrid detailed model, demonstrating the effectiveness of the physical microgrid dynamic equivalent in representing the microgrid dynamic behavior, following multi-microgrid islanding and under load-following conditions upon multi-microgrid islanding. Using the available physical knowledge it was possible to select a proper model structure with physical representation, which can be easily integrated in dynamic simulation tools, being the domain of validity extended to represent the microgrid dynamic behavior with respect to the MV network with the required accuracy.

5.7 Conclusions

The definition of an effective control scheme for multi-microgrid operation is a key issue in order to accommodate efficiently microgeneration and MV-connected DG units. In particular, it is extremely important to specify the interactions among controllers, and namely the CAMC that plays a central role in this control architecture.

The control strategy to be adopted should then be based on a hierarchical scheme, ensuring both autonomy and redundancy. A communication strategy for multi-microgrid systems is also important and must be capable of allowing an exchange of information able to handle properly the dimension and complexity of the system. Furthermore, the control strategy must take into account the two possible operating modes in a multi-microgrid system: (a) Grid-connected and (b) Islanded mode.

The main functions to be included in the CAMC, housed at the HV/MV substation level, should be adapted and/or duplicated from the traditional DMS modules. The functions identified in this chapter were: Coordinated voltage var support, coordinated frequency support and emergency functions (black start). Advanced voltage control functionalities, as the ones described in this chapter, will enable large scale integration of DG and micro-generation without jeopardizing the grid operating conditions and optimizing its performance.

Regarding the multi-microgrid stability assessment, dynamic and transient studies are required focusing on the multi-microgrid islanded mode of operation. Multi-microgrid islanding and load-following will play key role in these studies. Since performing these studies using detailed modeling approaches will be not be practical, microgrid dynamic equivalents have been developed in order to speed up time domain simulations. For this purpose, system identification based techniques exploiting TDNN and physical modeling can be used. From the studies performed we can deduce that physical microgrid dynamic equivalents are well fitted for this purpose.

The main issues to be dealt with in the future are the deployment of smart metering infrastructure as a means of pushing forward the development of microgrids as an integrated part of the general smart grid concept. The costs of developing the microgrid and multi-microgrid concepts appear to be large and their economic feasibility needs to be carefully evaluated. However, if these concepts are integrated within a smart metering deployment, to be used to manage commercially the trade of electricity (either consumed or generated) it will become easier to get an early successful deployment. This requires that an advanced metering infrastructure should be conceived in order to accommodate the additional communication requirement that will allow the control and management of the microgrid.

In addition, future challenges should include tackling the issues resulting from massive integration of distributed storage devices. Distributed storage technologies can be based on mobile storage such as electric vehicles or on stationary storage, which includes fuel cells, regenerative fuel cells and lithium-ion batteries. The full deployment of the microgrid and multi-microgrid concepts requires the development of distributed storage technologies and the assessment of their impact concerning technical, economic and regulatory issues. Complementarily, a full assessment of active demand-side management strategies must also be carried out.

Developing the smart grid concept means increasing flexibility for generation, consumption and grid management. Microgrids are a way of bringing additional flexibility to the system dealing with these three components. Also very important is the need to define the rules for this flexibility, which can include contractual supply of this capability (centralized/bilateral, voluntary/mandatory). This means that appropriate regulation and adequate solutions need to be sought for and developed within this field.

References

1. European Project "Advanced Architectures and Control Concepts for More Microgrids – More Microgrids", Project Reference no. 19864. Available at: http://www.smartgrids.eu/node/14.
2. Lopes, J.A.P. et al. (2003) "Management of Microgrids", presented at the International Electric Equipment Conference, Bilbao, Spain.

3. Lopes, J.A.P., Moreira, C.L., and Madureira, A.G. (2006) Defining control strategies for microgrids islanded operation. *IEEE T. Power Syst.*, **21**, 916–924.

4. Madureira, A.G., Pereira, J.C., Gil, N.J. *et al.* (2011) Advanced control and management functionalities for multi-microgrids. *Eur. T. Electr. Power*, **21** (2), 1159–1177.

5. Resende, F.O., Gil, N.J., and Lopes, J.A.P. (2011) Service restoration on distribution systems using multi-microgrids. *Eur. T. Electr. Power*, **21** (2), 1327–1342.

6. Masters, C.L. (2002) Voltage rise – The big issue when connecting embedded generation to long 11kV overhead lines. *Power Eng. J.*, **16** (1), 5–12.

7. Madureira, A.G. and Lopes, J.A.P. (2009) Coordinated voltage support in distribution networks with distributed generation and microgrids. *IET Renewable Power Generation*, **3** (4), 439–454.

8. Madureira, A. (2012) *Coordinated Voltage Control in multi-microgrids*, LAP Lambert Academic Publishing, Saarbrucken.

9. Schwefel, H.P. (1995) *Evolution and Optimum Seeking*, John Wiley & Sons, New York.

10. Kennedy, J. and Eberhart, R. (1995) "Particle swarm optimization", in Proc. IEEE International Conference on Neural Networks, vol. 4, Perth, Australia, pp. 1942–1948.

11. Miranda, V. and Fonseca, N. (2002) "EPSO – Best-of-two-worlds meta-heuristic applied to power system problems", in Proc. 2002 Congress on Evolutionary Computation, Honolulu, Hawaii, pp. 1080–1085.

12. Miranda, V. and Fonseca, N. (2002) "EPSO – Evolutionary Particle Swarm Optimization, a new algorithm with applications in power systems", in Proc. 2002 Asia Pacific IEEE/PES Transmission and Distribution Conference and Exhibition 2002, vol. 2, Yokohama, Japan, pp. 745–750.

13. Ciric, R.M., Feltrin, A.P., and Ochoa, L.F. (2003) Power flow in four-wire distribution networks – General approach. *IEEE T. Power Syst.*, **18** (4), 1283–1290.

14. Kundur, P. (1994) *Power System Stability and Control*, McGraw-Hill, New York.

15. Gil, N.J. and Lopes, J.A.P. (2007) "Hierarchical Frequency Control Scheme for Islanded Multi-Microgrids Operation", in Proc. IEEE PES PowerTech 2007, Lausanne, Switzerland.

16. Gil, N.J. and Lopes, J.A.P. (2008) "Exploiting Automated Demand Response, Generation and Storage Capabilities for Hierarchical Frequency Control in Islanded multi-microgrids", in Proc. 16th Power Systems Computation Conference, Glasgow, Scotland.

17. Microgrids Project Deliverable DD1 (2009), "Emergency strategies and algorithms", J.A. Peças Lopes. Available at: http://microgrids.power.ece.ntua.gr/micro/micro200/deliverables/Deliverable_DD1.pdf.

18. Moreira, C.L., Resende, F.O., and Lopes, J.A.P. (2007) Using low voltage microgrids for service restoration. *IEEE T. Power Syst.*, **22** (1), 395–403.

19. Lopes, J.A.P., Moreira, C.L., and Resende, F.O. (2005) "Microgrids black-start and islanding operation", in Proc. 15th Power Systems Computation Conference, Liège, Belgium.

20. Milano, F. and Srivastava, K. (2009) Dynamic REI equivalents for short circuit and transient stability analyses. *Electr. Pow. Syst. Res.*, **79** (6), 878–887.

21. Ramirez, J.M., Hernández, B.V., and Correa, R.E. (2012) Dynamic equivalence by an optimal strategy. *Electr. Pow. Syst. Res.*, **84** (1), 58–64.

22. Undrill, J. and Turner, A. (1971) Construction of power system electromechanical equivalents by modal analysis. *IEEE T. Power Ap. Syst.*, **PAS-90** (5), 2049–2059.

23. Marinescu, B., Mallem, B., and Rouco, L. (2010) Large-scale power system dynamic equivalents based on standard and border synchrony. *IEEE T. Power Syst.*, **25** (4), 1873–1882.

24. Miah, A.M. (2011) Study of a coherency-based simple dynamic equivalent for transient stability assessment. *IET Generation, Transmission & Distribution*, **5** (4), 405–416.

25. Podmore, R. and Germond, A. (April 1977) "Development of dynamic equivalents for transient stability studies", Systems Control, Inc. Technical Report.

26. Pires de Souza, E.J.S. (2008) Identification of coherent generators considering the electrical proximity for drastic dynamic equivalents. *Electr. Pow. Syst. Res.*, **78** (7), 1169–1174.

27. DeTuglie, E., Guida, L., Torelli, F. *et al.* (2004) "Identification of dynamic voltage-current power system equivalents through artificial neural networks", in Proc. Bulk Power System Dynamics and Control – VI, Cortina d'Ampezzo, Italy.

28. Stankovic, A.M. and Saric, A.T (2003) "An integrative approach to transient power system analysis with standard and ANN-based dynamic models", in Proc. 2003 IEEE Bologna PowerTech Conference, Bologna, Italy.

29. Shakouri, G.H. and Hamid, R.R. (2009) Identification of a continuous time nonlinear state space model for the external power system dynamic equivalent by neural networks. *Int. J. Elec. Power*, **31** (7–8), 334–344.

30. Azmy, A.M., Erlich, I., and Sowa, P. (2004) Artificial neural network-based dynamic equivalents for distribution systems containing active sources. *Proc. IEE Proceedings – Generation. Transmission and Distribution*, **151** (6), 681–688.

31. Azmy, A.M. and Erlich, I. (2004) "Identification of Dynamic Equivalents for Distribution Power Networks using Recurrent ANNs", in Proc. IEEE PES Power Systems Conference and Exposition, vol. 1, New York City, USA, pp. 348–353.

32. Resende, F.O., Moreira, C.L., and Peças Lopes, J.A. (2006) "Identification of Dynamic Equivalents for Microgrids with High Penetration of Solar Energy using ANNs," in Proc. 3rd European Conference of PV-Hybrid and Mini-Grid, Aix en Provence, France.

33. Resende, F.O. and Lopes, J.A.P. (2007) "Development of Dynamic Equivalents for Microgrids using System Identification Theory", in Proc. 2007 IEEE Power Tech, Lausanne, Switzerland, pp. 1033–1038.

34. Resende, F.O. (2007) "Contributions for Microgrids Dynamic Modelling and Operation", PhD Thesis, Faculty of Engineering, Porto University, Porto.

35. Barsali, S., Ceraolo, M., Pelacchi, P., and Poli, D. (2002) "Control techniques of Dispersed Generators to improve the continuity of electricity supply", in Proc. 2002 IEEE Power Engineering Society Winter Meeting, vol. 2, New York, USA, pp. 789–794.

36. Caldon, R., Rossetto, F., and Turri, R. (2003) "Analysis of dynamic performance of dispersed generation connected through inverter to distribution networks", in Proc. 17th International Conference on Electricity Distribution – CIRED, Barcelona, Spain.

37. El-Sharkh, M.Y., Rahman, A., Alam, M.S. *et al.* (2004) Analysis of active and reactive power control of a stand-alone PEM fuel cell power plant. *IEEE T. Power Syst.*, **19** (4), 2022–2028.

38. Katiraei, F., Iravani, M.R., and Lehn, P.W. (2005) Micro-grid autonomous operation during and subsequent to islanding process. *IEEE T. Power Deliver.*, **20** (1), 248–257.

39. Nelles, O. (2001) Nonlinear system identification, in *From Classical Approaches to Neural Networks and Fuzzy Models*, Springer Inc., New York.

40. Akagi, H., Kanazawa, Y., and Nabae, A. (1984) Instantaneous reactive power compensators comprising switching devices without energy storage components. *IEEE T. Ind. Appl.*, **IA-20** (3), 625–630.

6

Pilot Sites: Success Stories and Learnt Lessons

George Kariniotakis, Aris Dimeas and Frank Van Overbeeke (Sections 6.1, 6.2)

6.1 Introduction

The operation of microgrids offers the possibility of coordinating distributed resources in a more or less decentralized way, so that they behave as a single producer or load in energy markets. In this way, the full benefits of distributed resources can be exploited in a consistent, manageable way. A method for a systematic analysis of these benefits including a quantified evaluation is provided in Chapter 7, but a number of real-world microgrids are already in operation worldwide [1,2] as off-grid applications, pilot cases and full-scale demonstrations. Significant experience and various lessons can be already learnt from these real-world operating microgrids of all types, that is, community/utility, commercial, industrial, institutional, campus, military and remote off-grid, including pilot projects and commercial scale demonstrations.

The aim of this chapter is to provide a non-exhaustive overview of real-world microgrids currently in operation across the world, specifically in Europe, the USA, Japan, China and South America. Many of these demonstrations have been presented in a series of dedicated microgrid symposiums held in Berkeley, California, USA (2005), Mont Tremblant, Quebec, Canada (2006), Nagoya, Japan (2007), Kythnos Island, Greece (2008), San Diego, California, USA (2009), Vancouver, BC, Canada (2010), Jeju Island, Korea (2011), Évora, Portugal (2012), Santiago, Chile (2013). Presentations and other materials from these events are available at [3].

A number of equally interesting microgrid applications, such as in Bornholm island in Denmark, Jeju island in Korea and off-grid microgrids in Canada, are not included, since it was clearly impossible to cover all efforts in such a dynamically developing field in a single chapter.

6.2 Overview of Microgrid Projects in Europe

In the European Union (EU), microgrids form a key component in the Strategic Research Agenda for Europe's Electricity Networks of the Future [4]. In the past few years, substantial

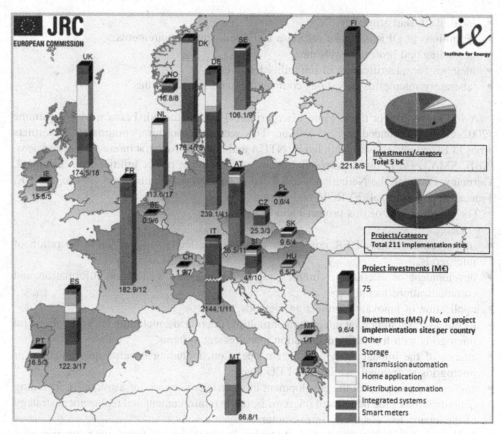

Figure 6.1 Geographical distribution of investments and project categories in EU. Source: JRC. Reproduced by permission of John Wiley & Sons Ltd

public and private investments have been made in R&D, demonstration and deployment activities in the smart grids area, including microgrids. Figure 6.1 shows a map of the 211 smart grid related projects running in EU27 at a total investment of €5 billion [5]. This amount includes more than €2 billion spent on smart meter rollouts. Besides the R&D activities in various projects related to microgrids, two major research efforts have been devoted exclusively to them. Within the 5th Framework Programme (1998–2002), the microgrid [6] activity was funded at €4.5 million. The Consortium, led by the National Technical University of Athens (NTUA), included 14 partners from seven EU countries, including utilities, such as EDF (France), PPC (Greece) and EdP (Portugal); manufacturers, such as EmForce, SMA, Germanos, Urenco and research institutions and universities such as Labein, INESC Porto, the University of Manchester, ISET Kassel and École de Mines. This project was successfully completed, providing several innovative technical solutions, including the development of:

- DER models plus steady-state and dynamic analysis tools enabling simulation of LV asymmetrical, inverter dominated microgrid performance,
- islanded and interconnected operating philosophies,
- control algorithms, both hierarchical and distributed (agent based),

- local black start strategies,
- definitions of DER interface response and intelligence requirements,
- grounding and protection schemes,
- methods for quantification of reliability benefits,
- laboratory microgrids of various complexities and functionalities.

A follow-up project titled More Microgrids [7] within the 6th Framework Programme (2002–2006) was funded at €8.5 million. This second consortium, comprising 22 partners from 11 EU countries, was again led by NTUA and included manufacturers such as Siemens, ABB, SMA, ZIV, I-Power, Anco, Germanos and EMforce; power utilities from Denmark, Germany, Portugal, the Netherlands and Poland, and research teams from Greece, the UK, France, Spain, Portugal, FYROM and Germany.

The achievements of this project include:

- investigation of new DER controllers to provide effective and efficient operation of microgrids,
- development of alternative control strategies using next generation information and communications technology,
- application of innovative protection methods,
- technical and commercial integration of multiple microgrids, including interface of several microgrids with the utility distribution management systems,
- studies of the impact on power system operation, including benefits quantification of microgrids at regional, national and EU levels,
- studies of the impact on the development of electricity network infrastructures, including quantification of the benefits of microgrids, to the reinforcement and replacement strategy of the ageing EU electricity infrastructure,
- field trials of alternative control strategies in actual installations, with experimental validation of various microgrid architectures in interconnected and islanded modes, and during transitions, testing of power electronics components and interfaces, and of alternative control strategies, communication protocols.

The change of scale from laboratories to real world pilot sites, realized by the More Microgrids project, has been a crucial step in drawing practical conclusions on different aspects of the operation and management of microgrids. The pilot sites include rural, residential, industrial and commercial microgrids in several countries. Two large laboratories were also used as pilot sites. This enabled field tests, which would have been risky to carry out on real-world microgrids. The Table 6.1 and Figure 6.2 give an overview of the pilot sites.

The following sections provide a description of the first three pilot microgrids mentioned in Table 6.1, the experimentation objectives at each pilot site, an overview of the results and some conclusions and lessons learnt.

6.2.1   Field Test in Gaidouromandra, Kythnos Microgrid (Greece): Decentralized, Intelligent Load Control in an Isolated System

The pilot microgrid electrifies an isolated settlement of 12 houses in a small valley of Kythnos, an island in the Aegean Sea, Greece (Figure 6.3).

Table 6.1 European Pilot Microgrids

No	Site (responsible partner)	Country	Type of microgrid
1	Gaidouromantra on Kythnos Island (CRES)	Greece	residential/island
2	Mannheim-Wallstadt settlement (MVV)	Germany	residential
3	Bronsbergen holiday park (Continuon/EMforce)	Netherlands	residential
4	Ilhavo municipal swimming pool (EDP Distribution)	Portugal	commercial
5	Bornholm Island (DTU, OESTKRAFT)	Denmark	multi-microgrid/island
6	AGRIA farm (UKIM, BIG)	FYROM	rural – commercial
7	LABEIN test facility (LABEIN)	Spain	large-scale test facility
8	ERSE test facility (ERSE)	Italy	large-scale test facility

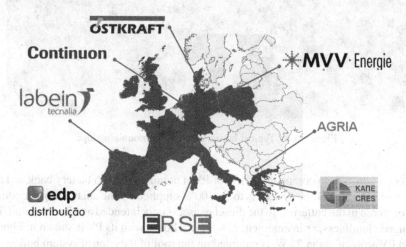

Figure 6.2 Pilot sites considered in the More-Microgrids project for field tests

Figure 6.3 Wide area view of the Gaidouromantra microgrid in Kythnos island

Figure 6.4 Typical house with PV on mountain slope

The generation system comprises 10 kW of PV, a nominal 53 kWh battery bank and a 5 kW diesel genset. The aim of the system is to be 100% supplied by the solar energy produced by the PVs or stored in the batteries, so the diesel genset is only intended as a back-up unit, in case of prolonged cloudiness, or in emergency. A typical house with its PV is shown in Figure 6.4. A second PV array of about 2 kW, is mounted on the roof of the control system building. This is a house of about 30 m^2 surface area, built in the middle of the settlement in order to house the battery inverters, the battery banks, the diesel genset and its fuel tank, the computer equipment for monitoring and the communication hardware, referred to as the system hut or system house (Figure 6.5). The 2 kW PV is connected to an SMA inverter and, along with a 32 kWh battery bank, they provide power for monitoring and communication. Residential service is powered by three Sunny Island battery inverters, connected in a parallel master–slave configuration, forming one strong single-phase circuit. This allows the use of more than one battery inverter only when more power is demanded by the consumers. Each battery inverter (SI4500) has a maximum power output of 3.6 kW. The battery inverters in the Kythnos system have the capability of operating in either isochronous or droop mode. The operation in frequency droop mode gives the possibility of passing information on to switching load controllers, in case the battery state of charge is low, and also to limit the power output of the PV inverters when the battery bank is full.

The electric system in Gaidouromantra is composed of the overhead power lines, a communication cable running in parallel to provide the monitoring and control needs. The grid and safety specifications for the house connections respect the technical solutions of the public power corporation (PPC), which is the local electricity utility. This decision was taken on the grounds that, potentially, the microgrid may be connected to the rest of the island grid.

Figure 6.5 The interior of the system house

The houses are supplied with single-phase electricity service, limited by a 6 A fuse. This means that each home can have lighting, a refrigerator, a water pump and some small electrical appliances. The residents were asked from the beginning to use high-performance appliances, such as fluorescent lamps and refrigerators with good insulation.

6.2.1.1 Objectives of the Demonstration

The primary goal of this demonstration was to test the decentralized control approach [8–10] described in Section 2.8. An agent-based load controller was designed and used to monitor the status of the power installation by taking measurements of voltage, current and frequency, in order to coordinate the energy management of the microgrid.

The objectives can be distinguished in technical and energy goals:

- The technical goal is to test a quite complex control system in a real environment. The Java-based MAS system has been tested in a real environment and occupied households. This means that the complex multi-agent system with all the functionalities (JADE platform, ontology, yellow pages capabilities, negotiation algorithm, coordination algorithm) should be installed and run smoothly for a long period of time.
- The electrical goals aim at the increase of energy efficiency by minimization of diesel generator usage and the shift of load consumption during hours of PV production excess. The goal is to encourage the operation of the most energy consuming loads, when there is excess of RES energy, and to curtail them when there is not enough energy in the batteries.

The curtailment should be fair (equal) for all households, so the "greedy" houses should be curtailed first. A secondary goal of the demonstration is to maintain the comfort level in the settlement. Therefore, if a resident feels the need to operate the water pump, or another controllable appliance (load), the agents are required to interfere to the minimum possible extent with this decision.

6.2.1.2 Technical Installation

Each house in the Kythnos microgrid is equipped with a water pump (1–2 kW), which is used to supply water to the residents of the house. The water pump is typically the largest load in a house and is considered as non-critical. Therefore, in case of energy shortage, it should be disconnected if needed.

The main feature of the system is the intelligent load controller (ILC) that hosts the Java agents described in Chapter 2. The controller measures electrical values such as voltage, current and frequency, and can remotely control up to 256 power line communication (PLC) load switches connected at any place inside the house. Each ILC controls two PLC switches. The first PLC switch controls the water pump, while the second controls a power socket, and any load connected to it (e.g. air-conditioning unit). Also, the ILC features a Wi-Fi interface that enables it to wirelessly connect to a local area network (Figure 6.6). This eliminates the need of a data cabling infrastructure, and simplifies the installation of the units.

The core of the unit is an integrated computer module that runs the Windows CE 5.0 operating system. The integrated computer module is driven by the powerful Intel® XscaleTM PXA255 processor at 400 MHz and features 64 MB of RAM and 32 MB flash memory (instead of a hard disk drive). Thus, it is suitable for demanding applications.

The operating system supports the installation of a Java virtual machine, and an agent environment based on the JADE platform has been embedded in the controller. This way, along with the MGCC installed in the system house of the Kythnos microgrid, the first actual field test of multi-agent systems in a microgrid has been realized.

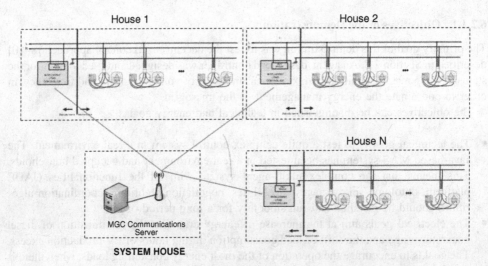

Figure 6.6 Communication among the load controllers

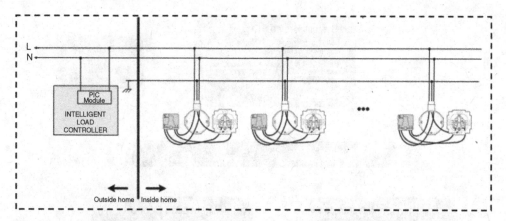

Figure 6.7 Connection of the load controller in the house electricity network. Reproduced by permission of the IEEE

The ILC hosts an integrated web server, through which it is feasible to remotely control the operation of each controller. During the installation in the Kythnos microgrid, a broadband internet connection has allowed remote control through a virtual private network. As a result, commands have been sent and data exchanged among the ILCs, the MGCC and authorized users who were not required to be onsite.

Figure 6.7 shows how the ILCs are connected in the house electricity network. Each ILC unit is connected to the power line outside the house, between the energy (kWh) meter and the house's electrical panel. The PLC switches are installed inside the electrical installation of the household.

Figure 6.8 shows pictures of the installed load controllers. For the installation of the ILCs, 5 houses were selected plus the System House where the battery inverters (Sunny Islands) are installed.

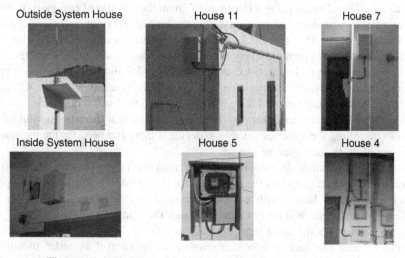

Figure 6.8 Installation of the load controllers outside the houses

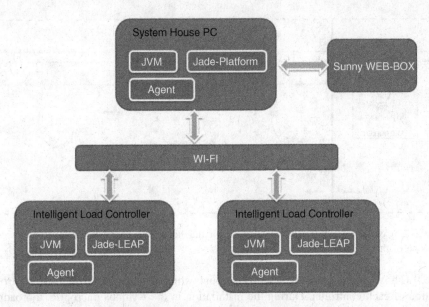

Figure 6.9 The agent-based software topology

6.2.1.3 Software Implementation

The key actors of the decentralized control system are the load agents and the MGCC.

Figure 6.9 presents the agent-based software topology. Each ILC hosts an agent. The microgrid central controller (MGCC) is housed in the system house. A PC hosts the agent platform and the software interfaces for communication with the Sunny web-box, a commercial product from SMA that hosts a web server and acts as a data logger for the PV or the battery inverters. These pieces of software form the multi-agent system.

The MGCC is responsible for monitoring the operation of the microgrid and coordinating the load agents. Therefore, it gathers information about the amount of energy consumed and the amount of energy produced respectively, from the load agents (intelligent load controllers) and the inverters of the PVs and the batteries. The MGCC is also informed about which controllable loads are in operation every moment. The controllable loads comprise the two loads in each house equipped with a PLC switch. If the production units of the microgrid (i.e. the PVs and the batteries) are able to supply the requested power, then the MGCC takes no action. If the PVs are capable of producing more power than the loads request, then the MGCC sends a message to the batteries informing the relevant agents that there is a surplus of power. The agents controlling the batteries are able to decide if there is a need for the batteries to be charged, according to their state of charge.

If the loads on operation demand more power than the production units can offer, the MGCC informs the load agents that there is a need for load shedding. The ILCs, equipped with intelligent agents, and hence with communication skills, negotiate in order to decide autonomously which load will be disconnected from the grid.

Figure 6.10 illustrates the main steps of the algorithm.

Both the PV and the battery inverters employ droop control in order to adjust and coordinate their operation, as described in detail in Chapter 3. This functionality needs to be

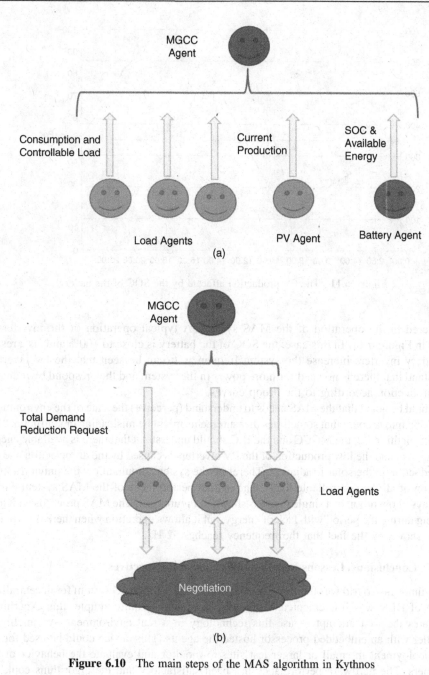

Figure 6.10 The main steps of the MAS algorithm in Kythnos

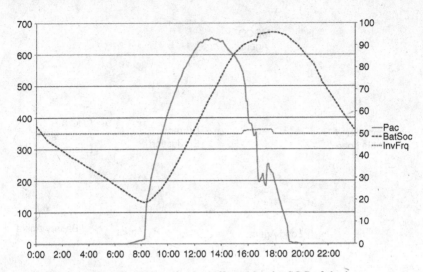

Figure 6.11 The PV production affected by the SOC of the battery

considered in the operation of the MAS system. A typical operation of the inverters is shown in Figure 6.11. In this case, the SOC of the battery is close to 100% and, as a result, the battery inverters increase the system frequency. It can be seen that the PV inverters understand that there is no need for more power in the system and they respond by reducing their production according to the droop curves.

It should be noted that the MAS needs to understand (perceive) the state of the environment, also taking into account that sometimes measurements might be misleading or conflicting. For example, in this case, the MGCC and the ILC should understand that there is available energy in the system and the low production of the PV inverters is caused by the droop action and not by a reduction in the solar irradiation. Therefore the system should allow the uninterruptible operation of all the controllable loads. The overall performance of the MAS system during three days of operation in a single house is shown in Figure 6.12. The MAS proceeds with load shedding during the period with lack of energy, but it allows operation when there is sufficient energy, shown by the fact that the frequency reaches 52 Hz.

6.2.1.4 Conclusions, Lessons Learnt and Technical Perspectives

The Kythnos microgrid was the first actual test site, where the MAS system for decentralized control of DER was implemented. Although the system is quite simple, this experiment constitutes the first attempt to use this technology in a real environment. An intelligent controller with an embedded processor hosted the agents. This device could be used for the MAS deployment in small or larger test sites to monitor and evaluate the behavior of the consumers. The hardware performance has been satisfactory and the algorithms could be processed fast enough. Minor technical problems concerning, for example, Wi-Fi signal losses due to the high levels of humidity, maloperation of the PLC switches, when the system frequency was above 52 Hz, could be solved with minor system modifications without affecting the citizens' comfort. It is clear that the MAS system developed is too complicated for the control application of just a few houses, and similar results could have been obtained by

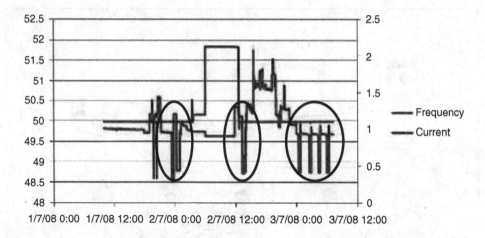

Figure 6.12 Load shedding in a specific house (indicated with ellipses). Reproduced by permission of the IEEE.

centralized control based on a simple set of "if . . . then . . . else" rules. The primary goal of the experiment, however, was to test techniques, such as negotiation algorithms, wireless communication and CIM-based ontology, for this application, and this has proven very successful. Moreover, further experiments have shown that the MAS technology offers great scalability for larger applications.

The main drawback of this technique is the high installation costs, including all the required electrical modifications in the internal electrical installations of the houses and the maintenance cost. This is one of the main debates in other EU projects. It has been shown that the knowhow required by the appropriate technical personnel should include ICT, as well as expertise in electrical installations. Furthermore, in some cases, the cost of the necessary electrical equipment required (electricity panels, switches etc.) can be comparable with, or even higher than, the cost of the load controller. This finding suggests that research efforts should further exploit internet technologies such Smart TV and smart appliances as described in Chapter 2.

6.2.1.5 Societal

Most of the citizens have accepted the system well and were very cooperative during the tests. They have allowed the installation of the intelligent controllers acting autonomously on their non-critical loads and they have accepted their cooperative behavior based on the equality principle of sharing the energy consumption. The MAS for energy optimization has provided the technical limitation and protection of the system to prevent overuse. This helped to maintain good relationships between the neighbours. The importance of involving or at least explaining to the users the operation of the system – and especially the negotiation process – cannot be overstressed. The graphical interface presented in Figure 6.13, has been used to demonstrate to the users the operational concept of the installed MAS in Kythnos. Clearly, in larger deployments, this task needs a dedicated effort undertaken by specialized personnel (customer service).

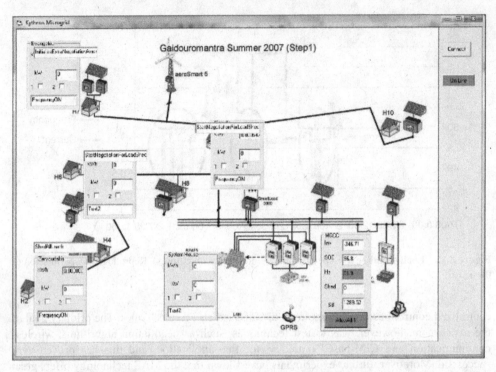

Figure 6.13 Simple electric diagram of the Kythnos microgrid

6.2.2 Field Test in Mannheim (Germany): Transition from Grid Connected to Islanded Mode

This field test site is an ecologically oriented residential estate in the Wallstadt district of Mannheim, Germany (Figure 6.14). It is considered as a key success factor for sustainable development to start awareness building with consumers and producers (pro-sumers) who have a positive attitude towards innovation and renewable energy. The selected settlement includes 580 households. The site includes several privately owned small photovoltaic systems (Figure 6.15) and one private Whispergen cogeneration unit. Further PV installations are in progress.

One measure for awareness building in preparation of the field test was the customization and installation of the display panel "VisiKid" at the entrance of the "Kinderhaus" in Mannheim-Wallstadt. VisiKid shows in real time, the current power level and how much energy has been produced by the PV panels (Figure 6.16). This intuitive explanation brings the timely value of energy from photovoltaic systems closer to parents and children.

The LV distribution grid represents a typical residential area with an intermeshed ring grid structure. It is fed from the MV grid via three 20/0.4 kV, 400 kVA transformers, and is operated in closed configuration. The LV grid, whose neutral is directly grounded at the MV/LV substation, is a three-phase network with distributed neutral. The LV voltage is 230/400 V (phase to neutral/phase to phase voltage) and the MV voltage is 20 kV. The apartments are connected to the electrical network by underground cables (Al $4 \times 150\,mm^2$). There are

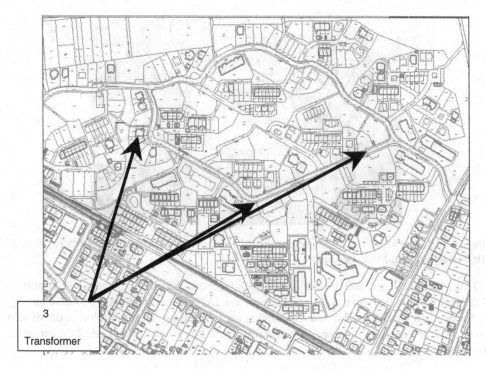

Figure 6.14 Map of the Wallstadt ecological estate. Courtesy: Dr. Frieder Schmitt, MVV, Germany

Figure 6.15 Typical house within the Wallstadt ecological estate. Courtesy: Dr. Frieder Schmitt, MVV, Germany

Figure 6.16 The VisiKid panel, displaying the PV generation at the Kinderhaus entrance. Courtesy: Dr. Frieder Schmitt, MVV, Germany

27 distribution boxes and further underground cables, which complete the LV grid in Wallstadt. Within the More Microgrids project, the LV grid was modified in order to prepare it for optional islanding. The new distribution area, including all relevant loads, distributed generators and storage, is served by one transformer. Power quality and grid characteristics have been monitored since 2006.

6.2.2.1 Description of the Experiments and Evaluation Results

For awareness building, the socio-economic experiment "Washing with the sun" was successfully realized. This was used to evaluate the willingness of customers to change the usage of household appliances, such as washing machines, dishwashers and tumble-dryers, to promote the consumption of local and solar-produced electricity. In days of good solar production, a message was sent to the participants specifying a time interval with high solar production and recommending the use of appliances in this period. A total of 24 households participated and shifted their loads significantly.

Two sets of technical experiments have been achieved in Wallstadt:

- realization of seamless transition between grid connected and islanded mode,
- installations of software agents responsible for the management of loads, storage and generators.

Seamless Transition between Grid Connected and Islanded Mode
Within the grid segment just described, the team prepared the Kinderhaus to operate as a microgrid, comprising two PV systems with a Sunny-Backup System, controllable loads and adding sufficiently designed battery storage as buffer, able to supply 10 kW for 1 hour.

To prove the feasibility of an iterative switching from islanded to grid-connected mode and vice versa, an additional battery for stabilizing the frequency was installed, in combination with a system of inverters able to realize an islanded mode. This battery (48 V 105 Ah, 21 A for 5 h) has been used, in combination with a system for the single purpose of regulating the frequency, while switching to islanded mode.

The Sunny-Backup System was connected to part of the loads inside the Kinderhaus, and an islanded mode was flawlessly realized in two days. During the periods of islanded operation, the frequency increased from 50 to 52 Hz caused by the increased SOC of the batteries. After reconnection to the grid, the frequency quickly returns to the 50 Hz imposed by the dominance of the grid.

Multi-Agent System for Energy Management

These experiments focused on the management of loads exposed to fluctuating generation. A combination of the agents' management abilities and the necessary communication links between the agents and their corresponding generators and/or loads was provided by a power-line network. In the network, IP addresses were assigned for each load and generator involved. The experiments carried out have shown that it is possible to control flexible loads (e.g. air-conditioning units) by increasing or decreasing their total consumed power according to a defined percentage of photovoltaic power, virtually assigned to them.

The system implementation focused on the usage of the agent-based software as well the available equipment at the test site. The system architecture was designed taking into account the existing broadband power line (BPL) communication networks. Unlike in the Kythnos pilot site, distributed I/O modules were extensively used, and the agents were hosted on central PCs, so the agents were not situated near the loads. This is graphically shown in Figure 6.17.

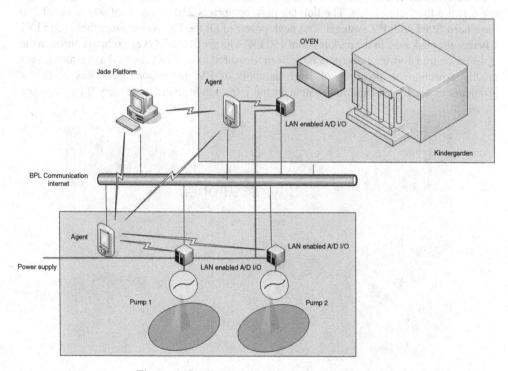

Figure 6.17 The system architecture at Wallstadt

6.2.2.2   Conclusions, Lessons Learnt and Perspectives

The main achievement of this experiment is the technical feasibility of the successful transition between the grid-connected and islanded mode of operation and vice versa in an automatic and seamless way. A further achievement is the increase of energy efficiency, that is, energy demand reduction as a result of the automatic negotiation between agents. These results are complementary, since during islanded operation an efficient demand-side management reduces load and thus helps to stabilize the islanded grid.

The ability to work in islanded mode increases the security of supply and thus contributes to the sustainability goals. This might be less important for strong, meshed grids, but could be well studied in Mannheim under real conditions without endangering the power supply to customers. DSOs operating weaker grids could be interested to develop grid segments as microgrids that are able to run in islanded mode during times of congestion or grid failures.

The microgrid installations in Wallstadt are further used for educational and demonstration purposes. The demonstration has been integrated into the educational concept of the Kinderhaus. Renewable energy and the timely value of electricity is now part of the daily routine and the annual curriculum for the children. Further schools in Mannheim would like to follow the visualization and experiments realized at this Kinderhaus.

6.2.3   The Bronsbergen Microgrid (Netherlands): Islanded Operation and Smart Storage

This demonstration case was set up at the Bronsbergen holiday park near Zutphen (Figure 6.18) in the Netherlands. The holiday park comprises 210 cottages, of which about 100 have been fitted with PV systems. The peak power of all the PV systems together is 315 kW, whereas the peak load in the park is about 150 kW. The profiles of power exchange through the distribution transformer in the park have been measured extensively, as well as various power quality aspects. The presence of large amounts of PV generation introduces particular problems in the LV distribution system in the park, among which are very high 11th and

Figure 6.18   Aerial view of the city of Zutphen, showing the Bronsbergen holiday park

13th harmonic currents and voltages. This situation, combined with the fact that, in high summer, the daily electricity consumption is more or less equal to the daily PV generation, made an ideal opportunity to explore the advantages of local storage, in combination with the improvement of power quality. This microgrid was therefore selected by Continuon (now Liander) for tests with "smart storage" and islanded operation.

A flexible AC distribution system (FADCS) has been proposed to address the following existing problems:

- Zero-sequence currents – third harmonic currents from the PV inverters cause considerable currents in the neutral conductor of the three-phase system, resulting in circulating currents on the delta side of the distribution transformer.
- Harmonics – the PV inverters inject considerable amounts of harmonics into the network. Also, the combined input capacitance of all inverters, together with the inductance of the distribution transformer, causes the system to show a poorly damped resonance at around 650 Hz. This resonance is excited heavily by the inverters.
- Asymmetry – the predominantly single-phase loads and generators result in poor symmetry of the three-phase voltages.

Moreover, the FADCS can be combined with electricity storage in batteries, which will support the envisaged islanded operation. Sunlight has supplied a total of 720 battery cells, each rated 2 V/500 Ah. The cells have been placed in four containers of 180 cells each. The smart storage and FACDS have been designed based on the microgrid concept.

The experimentation involved the contribution of several industrial actors:

- Liander, the local distribution network operator, acquired the test cottage (Figure 6.19) right next to the distribution transformer and made the battery containers, all the cabling and the distribution panels.

Figure 6.19 Outside view of the cottage used as the system house

- EMforce designed, manufactured and installed the inverters, the battery monitoring system and all control systems.
- Sunlight Systems manufactured and supplied the batteries.

6.2.3.1 Objectives of the Demonstration

Before any equipment was built, the three partners compiled a project plan outlining the following seven objectives:

1. *Islanded operation of the microgrid*
   During an average summer day, the microgrid should be capable of operating in islanded mode for a period of 24 hours. On a day with less than average irradiation from the sun, the microgrid may not be able to sustain 24 hours of islanded operation, but it should be able to survive 4 hours of islanded operation. In the Netherlands 4 hours is the maximum duration of a supply interruption for which a network operator is not required to pay a compensation to its customers.

2. *Automatic isolation and reconnection*
   The microgrid shall be equipped with an autonomous protection device which disconnects the microgrid from the external public utility network when a fault occurs on the external system. Furthermore, a system should be in place which is able to synchronize the microgrid to the external network once its voltage has returned, and to cause the interruption device to reconnect to the network. Internal faults must be handled selectively by the fuse on the faulted feeder. As a logical consequence of this approach, individual feeders have been installed for each of the storage systems. The fuses on these feeders will isolate the respective storage system in case of a fault within that system.

   Upon an enabling signal from the MGCC the protection device must reconnect the microgrid to the external network. For this purpose, the device shall be equipped with a synchronization detection facility. The FACDS must sense the voltage on the external network and adjust the voltage and frequency of the microgrid such that synchronous reconnection is possible. Reconnection must be established within one minute after it has detected that the voltage and frequency on the external network is within the statutory limits.

3. *Fault level of the microgrid*
   External faults: The short-circuit behavior of the FACDS should be such that it does not switch off during a fault on the incoming feeder from the public network, under the assumption that the protection system installed on that feeder operates as specified and isolates the microgrid from the feeder. Within 200 ms after the isolation is effective, the voltage within the microgrid should return to its normal value (i.e. 230 V within the tolerances as indicated in EN50166).

   Internal faults: The FACDS must be able to generate a fault current high enough to blow a 200 A fuse within 5 s. This corresponds to a minimum rms fault current of approx. 1200 A for 5 s. If the fault impedance is so low that the FACDS could not supply the fault current at its nominal output voltage, the voltage shall be lowered so as to limit the current to a value that can be sustained by the FACDS for 5 s. The aim here is to demonstrate that the islanded microgrid complies with the same safety standards as in grid-connected mode.

4. *Harmonic voltage distortion*

The FACDS should absorb and/or compensate harmonics within the microgrid to such an extent that each harmonic voltage from 100 Hz up to 2000 Hz complies with the planning level. The initial assumption for the planning level is: 1% of the rated 50 Hz voltage for each odd harmonic, except for the 9th, 15th, 21st, 27th, 39th; 0.2% of the rated 50 Hz voltage for even harmonics and for the 9th, 15th, 21st, 27th, 39th. If the harmonic level introduced by the public network would lead to very high harmonic currents between the public network and the FACDS, this requirement has to be verified in islanded mode only.

5. *Energy management and lifetime optimization of the storage system*

This involved the development and test of a so-called battery agent, which monitors the state of the battery and can issue requests for a regular full recharge to balance the individual cells and optimize the service life of the battery.

6. *Parallel operation of inverters*

For the evaluation of parallel operation of inverters, two identical FACDS have been built which are rated each for half the energy storage performance required. In order to be able to bridge short time periods with a single unit in operation, each unit shall be able to supply the full load current of the microgrid. Restrictions on the maximum power that can be absorbed by a single unit are acceptable.

The inverters shall not operate in master–slave mode; their control systems shall be designed such that parallel operation and a reasonable degree of load sharing is achieved without exchange of control signals between the two units.

7. *Black start of the microgrid*

The aim of this last objective is to demonstrate that the microgrid is capable of black starting from the battery.

6.2.3.2 Description of the Pilot Site

In order to be able to perform the experimentation mentioned, a circuit breaker was installed between the MV/LV transformer feeding the holiday park and the main LV busbar. Also, a number of fuses and contactors were added in order to provide sufficient flexibility in the various operating modes needed for the tests. The single-line diagram of Figure 6.20 illustrates these modifications. Two fuse groups were used to connect the inverters to the microgrid. If no testing was going on, the microgrid could be connected directly to the transformer so that all equipment, including the circuit breaker, could be isolated and safely accessible for staff.

The circuit breaker and the distribution board for the storage system were installed inside the system room (Figure 6.21), between the two inverter cubicles.

For reasons of size, weight and safety the batteries were installed in four purpose-designed containers, located in the "garden" of the cottage as illustrated in Figures 6.22 and 6.23.

The microgrid was controlled by an MGCC, which is designed to send setpoints to the two inverters, and contains the logic for automatic disconnection and reconnection of the microgrid. The circuit breaker was set so that if disconnection was caused by a fault current, it would trip automatically. If the microgrid was islanded as a consequence of an MV disconnection, an islanding test algorithm, implemented in the MGCC, would have detected that event and instructed the circuit breaker to open.

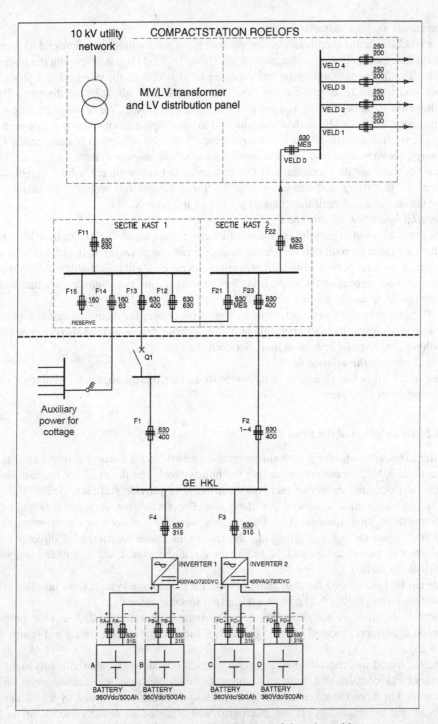

Figure 6.20 Schematic representation of the microgrid

Figure 6.21 The system room with switchgear cabinet (center) and inverters

The computer on which the MGCC was implemented also served as a platform for the battery monitoring system, which measured the voltage across each group of six battery cells. All measurements were logged on a second-by-second basis.

Each inverter was connected to a string of 360 cells connected in series, each rated at 2 V, 500 Ah. The rated voltage of a battery string was therefore 720 V.

Figure 6.22 View of the cottage and the battery containers. The distribution transformer is in the concrete box at the top right of the open container

Figure 6.23 Installation of the batteries

6.2.3.3 Experiments

All experiments served to confirm the achievement of one or more of the objectives already mentioned:

- *Parallel operation of inverters* – it was shown that in islanded mode, two inverters could be run in parallel in a stable manner without any master–slave function. Changes in active and reactive power were shared equally between the two inverters.
- *Fault level* – in islanded mode (but with the cottages connected directly to the transformer to avoid inconvenience to the owners) many short-circuit tests were done on a single inverter and on two inverters in parallel, both with a low-impedance fault and a fault including a simulated cable resistance. A typical example of recorded voltages and currents during a low-impedance phase-to-phase fault is shown in Figure 6.24. In all cases, the inverters generated enough current to blow the 200 A fuses installed in the distribution feeder within 5 seconds.
- *Battery management* – proper operation of the battery agent was demonstrated. It was used every few weeks for an automatic recharge of the batteries in order to optimize their service life. Many tests were done to establish the charging and discharging characteristics of the batteries and to estimate the energy efficiency of the storage system over a simulated 24-hour cycle.
- *Automatic isolation and reconnection, islanded mode* – these properties were tested in conjunction. Automatic isolation on an MV fault was demonstrated by simulation, as it was not permitted to test this with an evoked network fault. Automatic islanding detection and subsequent isolation was demonstrated successfully by testing.

The subsequent islanded mode was not sustained for 24 hours because the margin between available charge in the batteries and the energy consumed during the night appeared to be smaller than anticipated on the basis of measurements made 2–3 years previously. Therefore the local network operator preferred to test only during daytime.

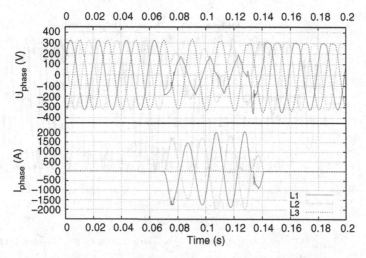

Figure 6.24 Typical example of a phase-to-phase short-circuit test

Both frequency and voltage stability during islanded mode were excellent and automatic reconnection to the external network has been done successfully.

- *Black start* – it was demonstrated repeatedly that a single inverter is able to black start the complete microgrid. A typical result is depicted in Figure 6.25. Within three (50 Hz) periods, the voltage has reached its normal value. After the standard synchronization time, the second inverter then connects to the network automatically and takes its share of the load.
- *Harmonics* – it was verified that the high 11th and 13th harmonic voltages are significantly reduced by the inverters. During the tests in islanded mode, it was found that the considerable rectifier loads in the microgrid caused high 3rd, 5th and 7th harmonic voltages. Therefore, additional active compensation has been added to the inverters, which reduced the harmonic levels in both grid-connected and islanded mode to values below the planned levels.

6.2.3.4 Conclusions – Lessons Learnt

- *Technical* – all technical systems used in the microgrid were developed on the basis of the architecture proposed in the More Microgrids project [7] and proved to be very appropriate for smooth operation in both grid-connected and islanded mode, as well as during the transitions between the two.

  Inverters for islanded operation must be able to actively control the harmonic voltage levels, because of the high percentage of electronic loads in residential networks. It was demonstrated that this is feasible.

  The benefit of having two identical parallel systems was shown, as it provides redundancy as well as an option to regenerate one of the batteries even in islanded operation.

  Islanding detection on a microgrid level was shown to be feasible but it is not as straightforward as it would be for a single inverter. This is supported by similar results from multi-inverter PV systems which were unable to detect an islanding condition of the whole

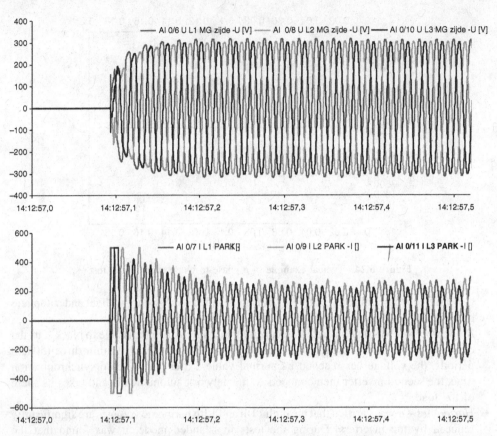

Figure 6.25 Black start: phase voltages (top) and phase currents (bottom) supplied to the microgrid at the instant of the black start

system. Further research in this area – as well as a proper definition of an islanding condition – is required to ensure safe operation of practical microgrids.

During the islanding tests, we found out that the PV inverters already installed in the park did not disconnect at a frequency above 52 Hz, which they should have done to comply with applicable grid codes. It has often been reported that commercial converters are not always compliant with grid codes. This is an issue of concern when planning multi-inverter systems and should be addressed by standardization or even legislation.

- *Economic* – the cost of a system as built in Bronsbergen is too high from the perspective of economic feasibility. In particular, the batteries are very costly and have a limited service life. This must also be weighed against the very high availability of the public utility system in the Netherlands. In countries or regions with poor standards of electric reliability, the economic argument could be different.
- *Operational* – the most important lessons learned are related to the way the operation of microgrids has to be introduced into the operational systems of a distribution network operator. These new technologies require additional education of staff on management, planning and shopfloor level, as well as the development of a whole new set of operational

standards and procedures. This is an item which is recommended as an essential part of any subsequent demonstration project.

- *Environmental* – grid-connected storage is an excellent way to harness the power generated by the PV systems in the microgrid. This avoids the necessity of upgrading the MV/LV transformer to cope with the high power generation during summer days at low load.
- *Societal* – it has not been easy to generate interest for these activities with the tenants of the cottages in the holiday park, although the project is happening on their doorstep. We have been successful in reducing inconvenience to them because Liander had the opportunity to acquire one cottage next to the MV/LV transformer and use it as a system house. The only visible technical items were the battery containers, located in the cottage's garden, which were painted green to minimize their visibility.

However, tenants who did take interest in technology, upon invitation, took the opportunity to get a guided tour through the installation and showed keen interest.

References

1. Hatziargyriou, N., Asano, H., Iravani, R., and Marnay, Ch. (July/August 2007) Microgrids: An Overview of Ongoing Research, Development and Demonstration Projects. Nr. 4, IEEE Power&Energy Magazine, pp. 78–94, Vol. 5.
2. Kroposki, B., Lasseter, R. *et al.* (May–June 2008) Making microgrids work. Issue 3, Page(s): 40–53, IEEE Power and Energy Magazine, Vol. 6.
3. http://der.lbl.gov.
4. http://ec.europa.eu/research/energy/pdf/smartgrids_agenda_en.pdf.
5. JRC Report, "*Smart Grid projects in Europe: lessons learned and current developments*". EUR 24856 EN, 2011.
6. "*MICROGRIDS: Large Scale Integration of Microgeneration to Low Voltage Grids*", ENK5-CT-2002-00610. 2003–2005.
7. "*MORE MICROGRIDS: "Advanced Architectures and Control Concepts for More Microgrids", FP6, Contract no.: PL019864.* 2006–2009.
8. Chatzivasiliadis, S.J., Hatziargyriou, N.D., and Dimeas, A.L. (20–24 July 2008) *Development of an agent based intelligent control system of microgrids.* IEEE Power and Energy Society General Meeting.
9. Dimeas, A.L. and Hatziargyriou, N.D. (8–12 Nov. 2009) *Control Agents for Real Microgrids.* Curitiba, Brazil: 15th International Conference on Intelligent System Applications to Power Systems, ISAP'09.
10. Hatziargyriou, Nikos. (5–7 December 2011) Intelligent Microgrid Control. Manchester: IEEE Innovative Smart Grid Technologies (ISGT) Europe 2011.

6.3 Overview of Microgrid Projects in the USA
John Romankiewicz, Chris Marnay (Section 6.3)

In recent years, the USA has become a leader in microgrid demonstration and technology development. The Department of Defense's (DOD) flagship microgrid project called Smart Power Infrastructure Demonstration for Energy, Reliability and Security (SPIDERS) and the Department of Energy's (DOE) grants given to nine microgrid demonstration projects have generated significant activity in the space, while other efforts in standards (IEEE 1547), technology (CERTS) and software (DER-CAM) have filled in key developmental gaps in the microgrid sector. In the absence of a federal clean energy policy, most states have been pursuing some form of clean energy legislation, with some positive developments for microgrids as well. This summary outlines the main R&D programs and demonstration projects as well as key developments in standards, technology and software in the USA, followed by a description of nine RDSI projects and three additional projects (Maxwell Air Force Base, University of California, San Diego and the Aperture Center).

6.3.1 R&D Programs and Demonstration Projects

The two major R&D and demonstration programs going on currently in the USA are SPIDERS, co-run by DOE, DOD and Department of Homeland Security (DHS) and the Renewable and Distributed Systems Integration (RDSI) microgrid grants program, run by DOE. While the goal of the SPIDERS program is to address energy security and reliability concerns, the RDSI grants are primarily focused on increasing the use of distributed energy during peak load periods to prove the value of microgrids for utility load shedding.

The goal of SPIDERS is "to reduce the 'unacceptably high risk' of mission impact from an extended electric grid outage by developing the capability to maintain energy delivery for mission assurance." SPIDERS will seek to demonstrate the following technologies at specific military base locations:

- cyber-security of electric grid – virtual secure enclave
- smart grid technologies and applications – advanced metering infrastructure, substation and distribution automation, two-way communications and control
- secure microgrid generation and distribution – islanding control system, seamless grid synchronization
- integration of distributed and intermittent renewable sources – PV, wind, solar, fuel cell, biofuels
- demand-side management – automated load shedding, smart sockets
- redundant backup power systems – batteries, vehicle to grid, other fuel sources

The timeline for the project rollout can be seen in Figure 6.26, with three project sites planned for Hickham Air Force Base (Hawaii), Fort Carson (Colorado) and Camp Smith (Hawaii) – see the map of Figure 6.27. The total three-year budget for the project is $39.5 million. The preliminary design for the Hickam base has been completed, while conceptual designs for Fort Carson and Camp Smith are in progress, with Requests for Information having been issued to dozens of potential industry partners. Once the three demonstrations have been completed, DOD hopes to create a template for implementation across the armed forces, as well as working with the National Institute of Standards and Technology (NIST) on technology transfer for the commercial sector and national grid cyber security. These military projects were preceded by a microgrid project at the Maxwell Air Force Base.

The goal of DOE's RDSI project is to demonstrate at least 15% peak demand reduction on the distribution feeder or substation level through integrating DER, and demonstrate microgrids that can operate in both grid parallel and islanded modes. If these goals are met, then many co-benefits will be realized, including increased grid reliability, deferment of utility transmission and distribution (T&D) investment, increased customer energy efficiency and decreased carbon emissions. Nine projects were selected in 2008 as part of the RDSI project, which has a total DOE budget of $55 million, with the total value exceeding $100 million with

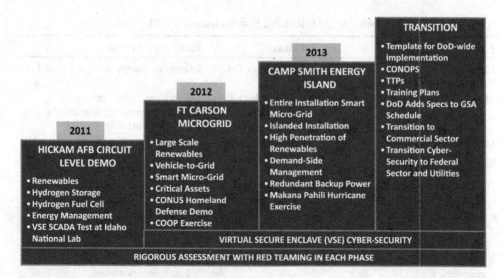

Figure 6.26 SPIDERS implementation plan [4]. Source: DOD 2011, contact: Jason Stamp, Sandia Labs. Sourced from DOD PPT: http://e2s2.ndia.org/schedule/Documents/2012%20Abstracts/Breakout%20Sessions/14191.pdf. Reproduced by permission of Jason Stamp

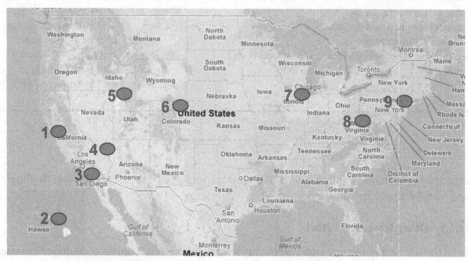

1. Santa Rita Green Jail
2. University of Hawaii
3. San Diego Gas and Electric
4. University of Nevada, Las Vegas
5. ATK Space Systems

6. City of Fort Collins
7. Illinois Institute of Technology
8. Allegheny Power West Virginia
9. ConEd New York

Figure 6.27 Map of DOE RDSI projects

Table 6.2 U.S. DOE Renewable and Distributed Systems Integration (RDSI) project list

Project lead and title	Location	Technologies/features
Chevron Energy Solutions – CERTS microgrid demonstration	Santa Rita Jail, CA	large-scale energy storage, PV, fuel cell
SDG&E– Beach Cities microgrid	Borrego Springs, CA	DR, storage, outage management system, automated distribution control, AMI
University of Hawaii – transmission congestion relief	Maui, HI	intermittency management system, DR, wind turbines, dynamic simulations modeling
University of Nevada Las Vegas – "Hybrid" Homes: dramatic residential demand reduction in the desert southwest	Las Vegas, NV	PV, advanced meters, in-home dashboard, automated DR, storage
ATK Space Systems – powering a defense company with renewables	Promontory, UT	hydro-turbines, compressed air storage, solar thermal, wind turbines, waste heat recovery system
City of Fort Collins – mixed distribution resources	Fort Collins, CO	PV, bio-fuel CHP, thermal storage, fuel cell, microturbines, PHEV, DR
Illinois Institute of Technology – the perfect power prototype	Chicago, IL	advanced meters, intelligent system controller, gas fired generators, DR controller, uninterruptable power supply, energy storage
Allegheny Power – WV super circuit demonstrating the reliability benefits of dynamic feeder reconfiguration	Morgantown, WV	biodiesel combustion engine, microturbine, PV, energy storage, advanced wireless communications, dynamic feeder reconfiguration
Con Ed – interoperability of DR resources	New York, NY	DR, PHEVs, fuel cell, combustion engines, intelligent islanding, dynamic reconfiguration and fault isolation

participant cost share. A full outline of the projects, including title, location and technologies used, can be found in Table 6.2.

6.3.1.1 Other Research Efforts

Given the strength of the research community among laboratories, universities and companies in the USA, a number of other research efforts in technology, software and standards have helped push the USA to the forefront of microgrid development.

The Consortium for Electric Reliability Technology Solutions (CERTS) runs a microgrid test bed facility in conjunction with American Electric Power in Groveport, OH, as well as a laboratory simulator at the University of Wisconsin, Madison. The consortium is currently focused on finding ways to accommodate intermittent distributed renewable energy sources within existing utility distribution systems and to find ways in which microgrids can seamlessly connect to and island from the grid. Currently, CERTS is adding multiple hardware units to its test bed facility, including:

- CERTS compatible conventional synchronous generator
- flexible energy management system for dispatch
- intelligent load shedding
- commercially available, stand-alone electricity storage device with CERTS controls
- PV emulator and inverter with CERTS controls.

CERTS technology has been applied in other microgrid projects around the USA, such as the Santa Rita Jail project led by Chevron Energy Solutions.

In the area of software development, the Distributed Energy Resources Customer Adoption Model (DER-CAM) has been developed by researchers at Lawrence Berkeley National Laboratory (LBNL) to predict and optimize the capacity, and minimize the cost of operating distributed generation and CHP for individual customer sites or microgrids. Based on specific site load (space heat, hot water, gas, cooling and electricity) and price information (electricity tariffs, fuel costs, operation and maintenance costs etc.), the model makes economic decisions on the distributed generation or CHP technologies that the user should adopt and how that technology should be operated. A schematic for DER-CAM can be seen in Figure 6.28. The model has been used internationally for about 10 years now.

Also in the area of software development, Energy Surety Micro-grids (ESM) is an assessment tool, with some similarities to DER-CAM and an eye towards assessing the possible application of microgrids at military bases. ESM uses a risk assessment methodology for the critical power delivery functions and needs of military bases, hospitals or communities. To assess the specific military applications, the US Army Engineer Research and Development Center (ERDC) has teamed up with Sandia National Laboratories to look at how microgrids can be implemented, not only at home bases, but also in field applications, such as forward base camps and tactical operation centers. So far, 12 bases across the USA have been evaluated, with more in the pipeline.

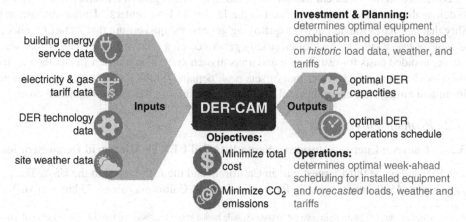

Figure 6.28 DER-CAM functionality

Finally, the USA has been a leader in common standards development for interconnection of DER to the grid, as well as islanding standards for microgrids. The development of these standards has streamlined business and safety operations for many projects and states trying to form regulations in these areas. In 2003, after five years of development, the IEEE 1547 *Standard for Interconnecting Distributed Resources with Electric Power Systems* was published with the goal of creating a unified technical requirement at a national level. In 2011, also after many years of drafting, the IEEE 1547.4 *Guide for Design, Operation and Integration of Distributed Resources Island Systems with Electric Power Systems* was published, covering microgrids and intentional islands. This standard is critical for the successful development of microgrids, and many state utility commissions will likely use it as principal guidance for any microgrid support policies they develop.

6.3.1.2 Policy Support

Although federal targets for renewable energy, distributed generation and microgrids are absent (Obama's administration has set a notional goal for 80% clean energy by 2035), federal policies have laid the groundwork for successful state policies in renewable energy and distributed generation. In 1978, the Public Utility Regulatory Policy Act (PURPA) requires that all electric utilities purchase all output of distributed generation projects and provide on-grid and backup services to all qualified distributed generation projects. The US Energy Policy Act of 2005 stipulated standard practice for net-metering and time-of-use metering. The Federal Energy Regulatory Commission (FERC) stipulated standards for interconnection of distributed energy generation projects (less than 20 MW) in 2006. Over the past decade, 44 states have established net-metering and interconnection policies, while 30 states have established renewable portfolio standards (RPS), although the strength of these net metering and RPS policies varies widely. A handful of states have specific carve-outs for distributed energy including Illinois, New Mexico and Arizona. Also, there are regional cap and trade programs running for the power sector in northeast USA (Regional Greenhouse Gas Initiative) and California (AB-32). In particular, California's cap and trade program may provide a promising environment for more development of CHP and microgrids. California municipal utilities have set tariffs for distributed renewable energy generators (not more than 3 MW) that are "strategically located and interconnected to the electrical transmission and distribution grid in a manner that optimizes the deliverability of electricity generated at the facility to load centers." Under this law, any utility that purchases the output from qualifying generators must ensure that its tariff "reflects the value of every kilowatt hour of electricity generated on a time-of-delivery basis, and shall consider avoided costs for distribution and transmission system upgrades, whether the facility generates electricity in a manner that offsets peak demand on the distribution circuit, and all current and anticipated environmental and greenhouse gases reduction compliance costs."

6.3.2 Project Summaries

6.3.2.1 Chevron Energy Solutions – Santa Rita Jail CERTS Microgrid Demonstration

Santa Rita Jail is the third-largest jail[1] in California and the fifth largest in the USA. The jail houses up to 4500 inmates and is located in Dublin, California, about 75 km east of San

---

[1] While the definitions are blurred, in general in the US, jails house prisoners awaiting trial or otherwise involved in the justice system, while prisons house convicted criminals.

Figure 6.29 Recent photo of Santa Rita Jail, with ground mounted tracking solar PV in the foreground and batteries and fuel cell directly behind

Francisco (Figure 6.29). Due to a series of installed DER and efficiency measures at the jail, it is often referred to as the "Green Jail." The aim of the microgrid project there is to demonstrate the first implementation of the CERTS microgrid technology, combined with large-scale energy storage, new and legacy renewable energy sources and a fuel cell. The goals as outlined by Alameda Country (the local county government in charge of the jail) are as follows:

- reduce peak electrical load and monthly demand charges
- store renewable and fuel cell energy overproduction
- shift electrical loads to off-peak hours
- improve grid reliability and reduce electrical voltage surges and spikes
- enable the jail to be a net-zero electrical facility during the most expensive summer peak hours
- expand the jail's onsite generation capacity to include three renewable energy sources: solar PV, wind turbines and solar water heaters.

Over the past decade, the project has implemented various energy efficiency measures and installed a wide array of distributed energy technologies, which have slowly accumulated into a full microgrid. In the spring of 2002, the jail installed a 1.2 MW rated rooftop PV array, followed in 2006 by a 1 MW molten carbonate fuel cell (MCFC) with CHP capability. Most recently, with the aid of DOE and California Energy Commission (CEC) grant money, as well as funding and participation from industry partners Chevron Energy Solutions, Satcon Power Systems and Pacific Gas and Electric, the jail has gained full microgrid capabilities with the installation of a large 2 MW, 4 MWh lithium-iron-phosphate battery, an islanding switch and associated power electronic upgrades.

In addition to generation equipment, the jail has also implemented a series of building equipment retrofits to improve efficiency and reduce peak electricity demand. A T-8 lighting ballast retrofit completed in 2009 was estimated to save 225 kW from the peak power demand and 1.34 GWh of electricity annually. Second, the 2010 installation of induction lighting in day rooms will save 217 kW peak power and 1.55 GWh annually. These two measures represent a 15% savings in peak power at the jail. Also, a number of other efficiency improvements were implemented in the middle of the 1990s targeting HVAC systems, lighting, refrigeration and other end-uses, which altogether decreased peak demand by 912 kW. Among them, an upgraded chiller took an estimated 423 kW off the peak power demand. The jail also plans to install a roof-mounted solar-thermal system which, when operational, will provide 40% of its hot water needs.

Lawrence Berkeley National Laboratory (LBNL) has also been an active partner with the jail, using its DER-CAM model a number of times throughout the development of the microgrid, to analyze electricity and heat requirements and to develop plans for the jail to meet its needs at minimum cost. Also, battery chemistries were compared, operating schedules that minimized risk were developed and opportunities for participation in demand response and ancillary services markets are now being analyzed.

6.3.2.2 SDG&E – Borrego Springs Microgrid Technologies Used

Borrego Springs is a small residential community located 90 miles northeast of San Diego. The microgrid there can be described as a utility-scale microgrid with its primary partner being San Diego Gas and Electric (SDG&E) and its primary focus is on smart grid service delivery models (Figure 6.30). The goal of the project is to provide a proof-of-concept test as to how information technologies and distributed energy resources (solar PV and batteries primarily) can increase utility asset utilization and reliability. The community already had many rooftop solar PV systems installed and is a somewhat isolated area fed only by a single sub-transmission line. Islanding of the entire substation area is being explored.

The project's partners include Lockheed Martin, IBM, Advanced Energy Storage, Horizon Energy, Oracle, Motorola, Pacific Northwest National Laboratories and University of California, San Diego. DOE supported the project with $7.5 million of federal funding, with additional funding coming from SDG&E ($4.1 m), CEC ($2.8 m), and other partners ($0.8 m).

The total microgrid installed capacity will be about 4 MW, with the main technologies being two 1.8 MW diesel generators, a large 500 kW/1500 kWh battery at the substation, three smaller 50 kWh batteries, six 4 kW/8 kWh home energy storage units, about 700 kW of rooftop solar PV and 125 residential home area network systems. The project will also incorporate feeder automation system technologies (FAST), outage management systems and price driven load management.

6.3.2.3 University of Hawaii – Transmission Congestion Relief

Hawaii is an ideal site for microgrids with its high energy costs, heavy reliance on oil, significant reliability issues and abundant renewable energy resources (solar, wind, tide and wave). The University of Hawaii's microgrid project's main goal is to develop and demonstrate an integrated monitoring, communications, data base, applications, and decision

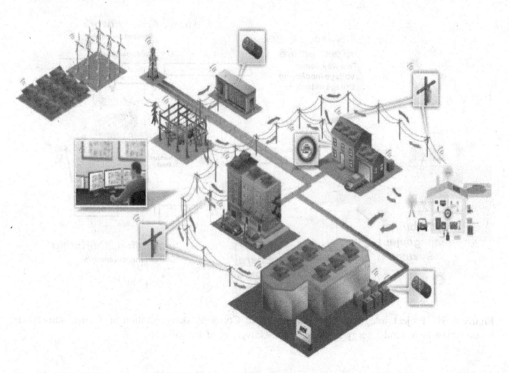

Figure 6.30 Schematic of SDG&E Borrego Springs microgrid project [17]. Source: Tom Bialek, SDG&E, 2011. http://energy.gov/sites/prod/files/EAC%20Presentation%20-%20SGD%26E's%20 Microgrid%20Activities%2010%202011%20Bialek.pdf. Reproduced by permission of Tom Bialek

support solution that aggregates distributed generation (DG), energy storage, and demand response technologies in a distribution system. The project will produce benefits for both the distribution and transmission systems, including reduced transmission system congestion, improved voltage regulation and power quality within distribution feeders, accommodation of the increase in distributed solar PV, better management of short-timescale intermittency from wind and solar energy, and provision for management of spinning reserve or load-following regulation. The Maui Electric Company (MECO) typically uses diesel-based generators for both conventional generation and reserves, but the growth in distributed generation – solar PV in particular – offers unique opportunities for microgrid applications (Figure 6.31).

The project is under the leadership of the Hawaii Natural Energy Institute (HNEI) of the University of Hawaii. The project team includes Maui Electric Company Limited (MECO), Hawaiian Electric Company, Inc. (HECO), Sentech (a division of SRA International, Inc.), Silver Spring Networks (SSN), Alstom Grid, Maui Economic Development Board (MEDB), University of Hawaii-Maui College (UH-Maui College), and the County of Maui. The project was supported by $7.0 million in funding from DOE and $9.0 million from the industry partners. Key technology focuses for the project include home area networks, advanced metering infrastructure, distribution management systems and distribution monitoring, and battery energy storage systems. The project has two phases. In Phase 1, energy management architecture for achieving project objectives will be developed and validated. In Phase 2, these capabilities will be demonstrated at a MECO substation at Wailea on Maui.

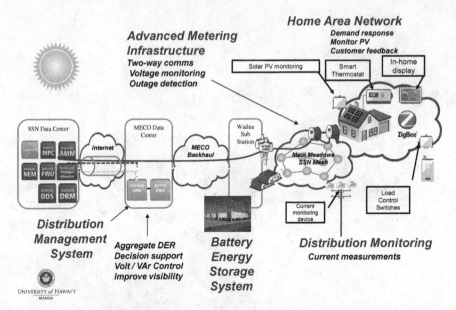

Figure 6.31 Project integration for Maui Electric Company demonstration at Wailea substation. Source: Hawaii Natural Energy Institute of the University of Hawaii [2]

6.3.2.4 University of Nevada Las Vegas – "Hybrid" Homes: Dramatic Residential Demand Reduction in the Desert Southwest

The RDSI project hosted by the University of Nevada Las Vegas (UNLV) is looking at dramatic demand reduction with highly energy-efficient homes that also use rooftop solar PV in a 185-unit housing development in northwest Las Vegas (Figure 6.32). The goal of the project is to decrease peak electrical demand for the new housing development by as much as 65% compared to basic codes. The homes, which are built to LEED platinum standards, utilize advanced metering technology, advanced wireless mesh network technology, Energy Star appliances, low-E windows, tankless water heaters and other smart grid and efficiency technologies. The project also incorporates battery storage at substations and small battery banks in several homes.

The homes' building partner, Pulte Homes, is working in conjunction with UNLV and Nevada Energy (local utility) on the project. They received $7.0 million in federal funding from DOE matched with $13.9 million in funds from the Pulte and Nevada Energy. For Nevada Energy, much of the funding came in the form of rebates through its Cool Share, Energy Plus and Zero Energy Home programs. Nevada Energy is also looking at dynamic pricing programs to make instantaneous power prices and cost incentives for demand reduction available to these homes.

6.3.2.5 ATK Space Systems – Powering a Defense Company with Renewables

ATK Space Systems is a large defense and aerospace manufacturing company in Magna, Utah, employing over 1500 people in 500–600 buildings (Figure 6.33). The main goal of the project was to develop and demonstrate a diverse system of renewable and storage

Figure 6.32 Rooftop solar PV systems that are a part of UNLVs RDSI energy efficiency homes project [2]

technologies that are integrated into an intelligent automation system with two-way communications to the utility to produce an on-demand reduction of 15% of substation load.

The main partners in the project include P&E Automation and Rocky Mountain Power (local utility). Funding includes $1.6 million from DOE and $2.0 million from industry

Figure 6.33 Birdseye view of ATK Space Systems [2]

partners. The peak demand of ATK Space Systems facilities is typically between 15–20 MW. The project already had 10 MW of on-site standby diesel generators for main grid event outages. This project's scope only includes a small portion of the company's total demand. Automated controls are expected to reduce the facility's electric demand by 3.4%, while battery storage and renewables will be dispatched during peak periods to avoid costs. The project will incorporate 100 kW of wind generation, 100 kW of waste heat generation and 1200 kWh of chemical-battery electrical storage.

6.3.2.6   City of Fort Collins – Mixed Distribution Resources

The Fort Collins Microgrid in Colorado is part of a larger project known as FortZED (Fort Collins Zero Energy District). The main goals are to develop and demonstrate a coordinated and integrated system of mixed distributed resources for the City of Fort Collins, reduce peak loads by 20–30% on multiple distribution feeders, increase the penetration of renewables and deliver improved efficiency and reliability to the grid and resource asset owners. The microgrid project is one of a kind as it involves multiple customers such as the New Belgium Brewery (Figure 6.34), InteGrid Laboratory, City of Fort Collins facilities, Larimer County facilities and Colorado State University, as well as various kinds of distributed energy generation technologies. Other technology partners in the project include Eaton, Advanced Energy and Brendle.

Technologies in the project include solar PV, CHP, microturbines, fuel cells, plug-in hybrid electric vehicles, thermal storage, load shedding and demand-side management. The peak demand of all customers is about 46 MW, while the combined distributed generation and load shedding capabilities is only 5 MW. The project has received $3.9 million in funding from DOE and $7.2 million from the various industry partners.

Figure 6.34   Rooftop solar PV installation on New Belgium Brewery

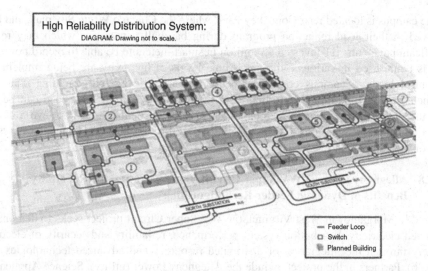

High Reliability Distribution System:
DIAGRAM: Drawing not to scale.

— Feeder Loop
☐ Switch
🔩 Planned Building

Figure 6.35 Diagram of IIT perfect power prototype [18]. Reproduced by permission of Mohammad Shahidehpour

6.3.2.7 Illinois Institute of Technology: Perfect Power Prototype

There have been a number of drivers for the Illinois Institute of Technology (IIT) to construct the perfect power prototype (Figure 6.35). First, the occurrence of at least three power outages per year resulted in a series of teaching and research disruptions with an estimated cost of $500 000 annually. The campus was also facing growing demand for energy, and the need to add infrastructure to accommodate its growth, update costly old infrastructure, improve energy efficiency and reduce consumption. IIT, in collaboration with the Galvin Electricity Initiative (GEI) and other key partners, is leading an effort to develop and validate innovative smart grid technologies and demonstrate smart grid applications, community outreach and renewed policies for better serving the consumers. This microgrid is sponsored by $7 million of federal funds (DOE) and $5 million of industrial funds over five years. Its main purpose and objectives are to create a self-healing, learning and self-aware smart grid that identifies and isolates faults, reroutes power to accommodate load changes and generation and dispatches generation and reduces demand based on price signals, weather forecasts and grid disruptions.

The IIT prototype will be the first of a kind of integrated microgrid system that provides for full islanding of the entire campus load, based on PJM[2]/ComEd market signals. Specific innovative technology applications include: high reliability distribution system, intelligent perfect power system controller, advanced ZigBee wireless technology, advanced distribution recovery systems, buried cable fault detection and mitigation.

The peak load of IIT's campus is around 10 MW. Their on-campus DER includes two 4 MW combined cycle gas units and a small wind turbine, with plans to add rooftop PV, as well as a 500 kWh battery. Total DER capacity would then be close to 9 MW, so the campus would be able to operate as an island most of the time, not importing any power from the grid. Full islanding capability has also been tested.

---

[2] PJM is the regional transmission organization (RTO) that Illinois is a part of.

The campus is located near Comiskey Park, where the Chicago White Sox play, and IIT is involved in their load reduction program during baseball games, for which they receive significant payments. IIT invested $3 million in smart meters to be able to record how much load is being used in various buildings. Three levels of hierarchy are being implemented to control loads: a campus controller (being internally developed), building controllers (Siemens) and sub-building controllers (ZigBee). Around 20% of IIT's load can be shed with the potential to reduce peak load by up to 50% on demand, and achieve a 4000 t/a reduction in carbon emissions. IIT has put out a request for proposal for demand response for 25% of the campus's total load.

6.3.2.8 Allegheny Power – WV Super Circuit, Demonstrating the Reliability Benefits of Dynamic Feeder Reconfiguration

Located in Morgantown, West Virginia, the WV Super Circuit project seeks to demonstrate improved electricity distribution system performance, reliability and security of electricity delivery through the integration of distributed resources and advanced technologies (Figure 6.36). Partners in the project include the Allegheny Power (utility), Science Applications International Corporation, West Virginia University Advanced Power and Electricity Research Center, Intergraph, North Carolina State University, Augusta Systems, First Energy and Tollgrade Communications. The project received $4.0 million in funds from DOE, matched with $5.4 million from industry.

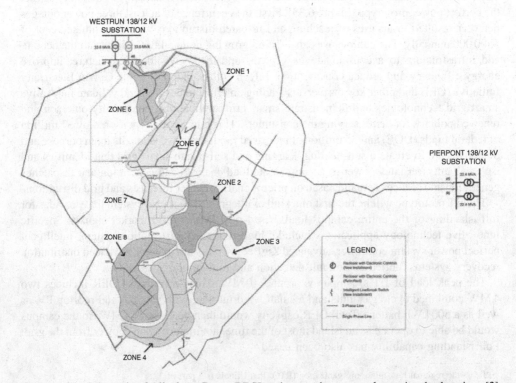

Figure 6.36 Schematic of Allegheny Power RDSI project service area and associated substations [2]

The microgrid load includes two commercial buildings with a combined load of about 200 kW for two tenants and distributed energy resources including 160 kW of natural gas internal combustion engine generators, 40 kW of solar PV and energy storage capable of providing about 24 kW for a two-hour period. A First Energy-controlled point of common coupling has an automatic switch to allow for intentional islanding of the buildings. Beyond the microgrid involving these two buildings, the project partners are also looking into the following technologies for improved grid operation: multi-agent grid management system, demand response and automated load control, low-cost distribution sensors, fault location and prediction and dynamic feeder reconfiguration.

6.3.2.9 ConEd – Interoperability of Demand Response Resources

This project run by the utility ConEd in New York City seeks to utilize a demand response command center to aggregate multiple DR resources at retail electric customer sites to supply critical services, under tariff-based and market-based programs, to the electric distribution company and to the regional transmission operator. Load reduction will be coordinated at 29 different telecom facilities and the ConEd headquarters into one "visualization platform". Distributed diesel generator resources totaling 20 MW on 24 customer sites will be aggregated as well (Figure 6.37). A mapping platform shows operators the status of different grid resources (customer distributed generation and demand response), using different colors to depict different operating status conditions. A key policy issue for this project will be how to price these demand response and distributed generation aggregation activities.

The project received $6.8 million in funding from DOE, matched with $6 million in funds from industry and $1 million from the New York State Energy Research and Development Authority. Partners include Verizon, Infotility and Innoventive Power.

6.3.2.10 Maxwell Air Force Base (A Military Microgrid)

The Maxwell AFB microgrid is a research and development project to validate the basic functionality of autonomous engine controls based on the CERTS droop control concept.

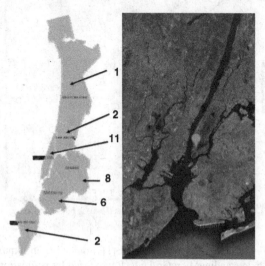

Figure 6.37 Numbers of customer site located throughout New York City for the ConEd RDSI project [2]

It does this by modifying the controls of existing diesel gensets, and operating those with new generation that is located some distance away from the existing gensets but on the same distribution feeder. The goal is to determine if these generators can share the loads and maintain stability in an islanded mode.

The specifics of the project include two 600 kW diesel backup gensets that are located in one building and installing a new CERTS-based 100 kW genset in a different building some distance from the first. The existing feeder connecting these two buildings will be sectionalized from the other loads by installing switchgear in the appropriate locations. This switchgear will isolate these two buildings from other loads on the feeder and create an experimental microgrid with two building loads and three generators.

Successfully demonstrating the stability of the controls will allow expansion of this microgrid to include more loads and additional generators that will maintain a stable microgrid, even in the absence of a central command and control architecture common to most microgrids today.

The ability to modify existing generators while adding new gensets with the CERTS droop functionality is an important milestone in the future deployment of microgrids, because a vast majority of existing buildings with mission-critical functions already have legacy backup gensets that still have ample operational life left in them. Also, integration of new gensets with CERTS controls and renewable generation sources with inverters that have similar functionality can be readily integrated into such microgrids.

6.3.2.11 UCSD Project (A Large Campus Microgrid)

The UCSD microgrid project supplies electricity, heating and cooling for a 450 hectare campus with a daily population of 45 000 (Figure 6.38). The main two motivations for constructing the project are as follows:

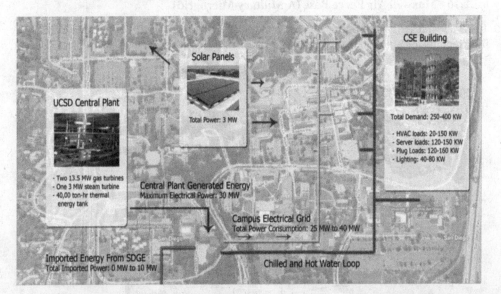

Figure 6.38 Energy flow through UCSD microgrid [1]. Sourced from http://synergy.ucsd.edu/files/ Agarwal_DATE2011_UnderstandingMicroGrid.pdf. Reproduced by permission of Yuvaj Agarwal

- After deregulation in California, the campus was able to purchase gas at an attractive rate to generate power by itself. For example, the CHP plant that they built had a five-year capital cost payback period based on avoided gas purchase costs.
- There is an existing campus steam distribution system for UCSD to have the ability to use steam to drive chilled water for cooling as well as hot water and heating.

The UCSD microgrid consists of two 13.5 MW gas turbines, one 3 MW steam turbine and a 1.2 MW solar-cell installation that together supply 85% of campus electricity needs, 95% of its heating and 95% of its cooling. The turbines produce 75% less emissions of criteria pollutants than a conventional gas power plant. For HVAC, it uses a 40 000 ton/hour, 3.8 million-gallon capacity thermal energy storage bank, plus three chillers driven by steam turbines and five chillers driven by electricity. A 2.8 MW molten carbonate fuel cell is running on waste methane, which is sponsored by California's self-generation incentive program funds and takes advantage of a 30% federal investment tax credit. The campus is connected to SDG&E by a single 69 kV substation. The UCSD uses a "straight SCADA system" for the building systems and energy supply to ensure their communication with each other. UCSD is installing a new high-end Paladin master controller, which will control all generation, storage and loads with hourly computing to optimize operating conditions. It can receive as many as 260 000 data inputs/second. To support Paladin, UCSD will use VPower software to process market-price signals, weather forecasts and the availability of resources. At the UCSD campus, about 200 power meters, on the main lines and at buildings' main circuit breakers, track use minute by minute. Lastly, DOE has just given USCD a grant to model the effects on the local distribution system from the ramping up and down of the solar PV system's output.

6.3.2.12 Aperture Center Project (A Green Field Commercial Building Microgrid)

The Aperture Center in Mesa del Sol, Albuquerque, New Mexico, will be the test site for a commercial microgrid. The project is a collaborative effort between the USA and Japan. It is being carried out by NEDO (see Section 6.4), along with the State of New Mexico, Mesa del Sol, Public Service Company of New Mexico, Sandia and Los Alamos National Lab and Los Alamos County.

This demonstration is bringing Japanese technology to demonstrate how to integrate multiple generation sources including renewable energy resources along with multiple storage sources, and how they can be optimized to interact with the building load. A number of Japanese companies are participating in the project, including Toshiba, Sharp, Fuji Electric, Tokyo Gas and Mitsubishi. Each of the Japanese companies has specific research that they want to do in this project. The Japanese have funneled about $30 million, statewide, into many smart grid projects in New Mexico, including this project at the Aperture Center. The local utility company is only involved in order to ensure that it interconnects and operates safely within the distribution grid.

The system comprises a 50 kW solar PV system (Figure 6.39) mounted on a shade structure over a parking lot and utility yard, currently under construction, that will contain an 80 kW fuel cell, a 240 kW natural gas-powered generator (Figure 6.40), a lead-acid storage battery power system and hot and cold thermal storage. All parts will be interconnected through a control room and building management system in the Aperture Center. The project is on schedule to be up and running in mid to late spring of 2012.

Figure 6.39 Natural gas generator at the Aperture Center microgrid

Figure 6.40 Solar PV array at the Aperture Center microgrid

References

1. Agarwal, Y., Weng, T., Gupta, R.K. 2011. "Understanding the Role of Buildings in a Smart Microgrid," http://synergy.ucsd.edu/files/Agarwal_DATE2011_UnderstandingMicroGrid.pdf.
2. Bossart, S. (2009) "Renewable and Distributed Systems Integration Demonstration Projects." EPRI Smart Grid Demonstration Advisory Meeting. http://www.smartgrid.epri.com/doc/15%20DOE%20RDSI%20Project%20Update.pdf.
3. DeForest, N., Lai, J., Stadler, M., Mendes, G., Marnay, C., and Donadee, J. (2012) *Integration and operation of a microgrid at Santa Rita Jail*, Lawrence Berkeley National Laboratory (LBNL-4850E).
4. DOD (Department of Defense) (2011) "Smart Power Infrastructure Demonstration for Energy Reliability and Security (SPIDERS)." http://www.ct-si.org/events/APCE2011/sld/pdf/89.pdf.
5. DSIRE (Database of State Incentives for Renewables and Efficiency) (2011) http://www.dsireusa.org/.
6. General Services Agency, County of Alameda (2012) "Alameda County Santa Rita Jail: Achievements in Energy and Water Efficiency." http://www.acgov.org/pdf/GSASmartGRIDProject.pdf.

7. Hatziargyriou, N., Asano, H., Iravani, R., and Marnay, C. (2007) Microgrids – an overview of ongoing research, development, and demonstration projects. IEEE Power & Energy Magazine, July/August 2007, pp. 78–94.
8. IEEE (Institute of Electrical and Electronic Engineers) (2003) "Standard for Interconnecting Distributed Resources with Electric Power Systems."
9. LBNL (Lawrence Berkeley National Laboratory) (2003) Integration of Distributed Energy Resources: the CERTS Microgrid Concept. Prepared for California Energy Commission. http://certs.lbl.gov/pdf/50829.pdf.
10. Lidula, N.W.A. and Rajapakse, A.D. (2011) Microgrids research: a review of experimental microgrids and test systems. Renewable and Sustainable Energy Reviews, 15, 186–200.
11. Marnay, C., Asano, H., Papathanassiou, S., and Strbac, G. (2008) Policymaking for Microgrids: Economic and Regulatory Issues of Microgrid Implementation. IEEE Power and Energy Magazine, May/June 2008, pp. 66–77.
12. Marnay, C., DeForest, N., Stadler, M. et al. (2011) A Green Prison: Santa Rita Jail Creeps Towards Zero Net Energy, Lawrence Berkeley National Laboratory, LBNL-4497E.
13. Navigant Consulting (2006) "Microgrids Research Assessment – Phase 2." http://der.lbl.gov/sites/der.lbl.gov/files/montreal_navigantmicrogridsfinalreport.pdf.
14. UCSD Sustainability Solutions Institute (2010) "Smart power generation at UCSD." http://ssi.ucsd.edu/index.php?option=com_content&view=article&id=416:smart-power-generation-at-ucsd-november-1-2010&cat-id=8:newsflash&Itemid=20.
15. Ustun, T., Ozansoy, C., and Zayegh, A. (2011) Recent developments in microgrids and example cases around the world-A review. Renewable and Sustainable Energy Reviews, 15, 4030–4040.
16. Young Morris, G., Abbey, C., Joos, G., and Marnay, C. (2011) A Framework for the Evaluation of the Cost and Benefits of Microgrids, Lawrence Berkeley National Laboratory (LBNL-5025E).
17. Bialek, T. (2011) "SDG&E's Microgrid Activities," DOE Electricity Advisory Committee. http://energy.gov/sites/prod/files/EAC%20Presentation%20-%20SGD%26E's%20Microgrid%20Activities%2010%202011%20Bialek.pdf.
18. Kelly, J., Meiners, M., and Rouse, G. (2007) "IIT Perfect Power Prototype: Final Report," Galvin Electricity Initiative and Illinois Institute of Technology. http://www.galvinpower.org/sites/default/files/documents/IIT_Perfect_Power_Prototype.pdf.

6.4 Overview of Japanese Microgrid Projects

Satoshi Morozumi (Section 6.4)

The New Energy and Industrial Technology Development Organization (NEDO) is Japan's largest public R&D management organization for promoting the development of advanced industrial, environmental, new energy and energy conservation technologies. One of the important objectives of NEDO's R&D is solving problems that arise when distributed and renewable resources are connected to power grids. These issues arise because the power output from most renewable energy resources fluctuates with weather conditions, and connecting them to traditional power grids may create power quality issues. Therefore, the development of energy management systems, energy storage applications and forecasting methods is important for resolving connection issues. NEDO is promoting several grid connection related projects, as shown in Figure 6.41. In those projects, two microgrid-related projects are involved. After the year 2010, NEDO started several international smart community projects shown in Figure 6.42. One of those projects, the New Mexico smart grid project, includes two distribution level microgrids and two customer level microgrids.

6.4.1 Regional Power Grids Project

The Demonstrative Project of Regional Power Grids with Various New Energies, undertaken from FY2003–FY2007, is one of the most notable projects in the history of NEDO. This

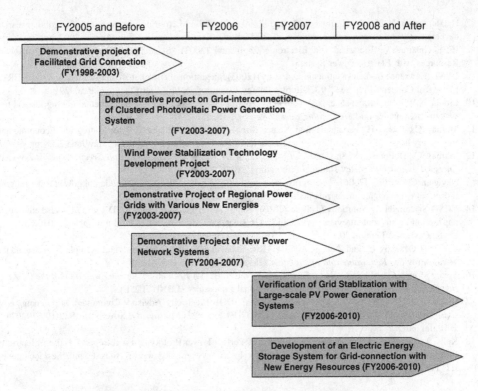

Figure 6.41 NEDO's grid-connection system projects

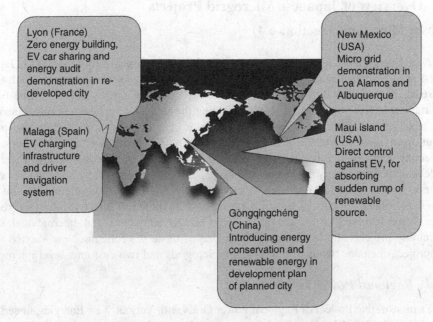

Figure 6.42 NEDO's international smart community projects

project encompasses three related sub-projects, one of which was conducted at the site of The 2005 World Exposition, Aichi, Japan.

6.4.1.1 Aichi Micro Grid Project

In the Aichi project, a power supply system utilizing fuel cells, photovoltaic cells and a battery system, all equipped with inverters, was constructed. A diagram of the supply system for the project is shown in Figure 6.43.

The primary power generation sources for this microgrid system were the fuel cells and the PV systems. The fuel cells included two molten carbonate fuel cells (MCFCs) with capacities of 270 kW and 300 kW, one 25 kW solid oxide fuel cell (SOFC) and four 200 kW phosphoric acid fuel cells (PAFCs). City gas was the primary fuel for the fuel cells. However, some of the fuel for the MCFCs was supplied by a methane fermentation system and a gasification system. The total capacity of the installed PV systems was 330 kW and multi-crystalline silicon, amorphous silicon and single crystalline silicon bifacial cells were used. Also, a sodium-sulfur (NaS) battery was used to store energy within the supply system and it played an important role in balancing supply and demand, as shown in Figure 6.44.

This demonstrative power plant (Figure 6.45) was installed at the site of the 2005 World Exposition, Aichi, Japan (EXPO 2005) and operated from December 2004 to September 2005. During the demonstration period, a total of 3 716 MWh of electricity was supplied by the power plant to two major pavilions. After EXPO 2005, the power plant was relocated to a site in Tokoname City, near the Chubu International Airport (Figure 6.46), and demonstrative operation was restarted in August 2006 and concluded in December 2007.

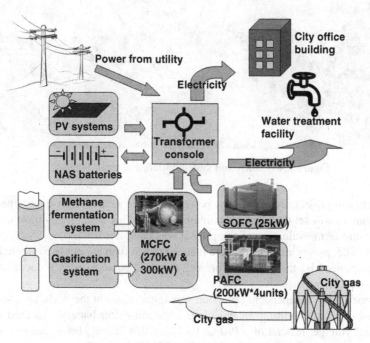

Figure 6.43 Diagram of the Aichi project

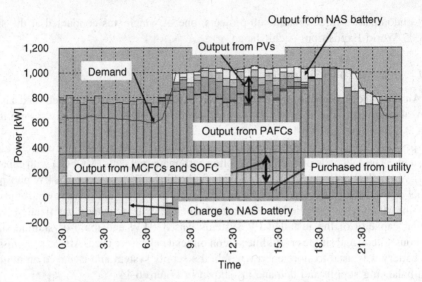

Figure 6.44 Typical daily operation of the Aichi microgrid

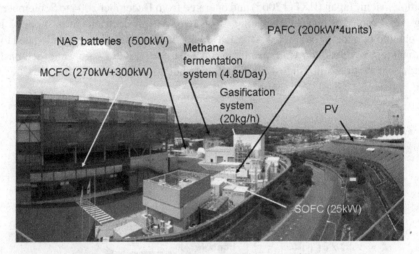

Figure 6.45 Microgrid system installed at the EXPO site

In the Aichi sub-project, since all of the power resources were inverter-based, the target for power imbalances was set to 3% per 30 minute interval. This target was almost achieved by the time the mid-term evaluation of the project was conducted.

During FY2005, power imbalances were reduced to less than 8% per five minute interval in the Kyotango project. By the end of the project, power imbalances were reduced to less than 3% per five minute interval.

In September 2007, independent operation was tested again at the Aichi site, for one day, under limited demand condition. In this case, a sodium-sulfur battery was used to control frequency and voltage, instead of a PAFC. By using this battery, better power quality was achieved relative to the first test conducted in 2005.

Figure 6.46 Microgrid system relocated near Chubu international airport

6.4.1.2 Kyotango Virtual Microgrid

The system installed in Kyotango was not technically a microgrid because each energy supply facility and demand site was connected to a utility grid and was only integrated by a control system. This energy supply system, as shown in Figure 6.47, was, more accurately, a "virtual microgrid." This concept is very useful in the area where grid system is fully deregulated.

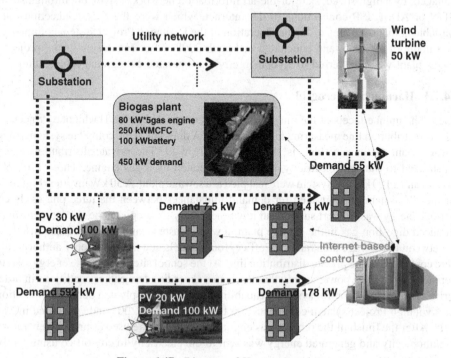

Figure 6.47 Diagram of Kyotango project

Figure 6.48 Kyotango biogas plant

The main facility of the system (Figure 6.48) is a biogas plant. Gas engines with a total capacity of 400 kW were installed in the plant, together with a 250 kW MCFC and a 100 kW lead-acid battery. In remote locations, two PV systems and a 50 kW wind turbine were also installed. The power generation equipment and end user demand were managed by remote monitoring and control. One of the interesting features of the system is that it was not managed. by a high-speed, state-of-the-art information network system, but through standard ISDN or ADSL ISP connections to the internet, which were the only connection options available in that rural area of Japan. Operation of the Kyotango Project system commenced in the middle of FY2005, and concluded in March 2008. After conclusion of the project, the biogas plant was transferred to Kyotango city and operated as regional refuse dump.

6.4.1.3 Hachinohe Microgrid

One of the unique facets of the microgrid system constructed in the Hachinohe Project was a private distribution line measuring more than 5 km. A diagram of the complete system, from the sewage plant to the city hall, is shown in Figure 6.49. The private distribution line was constructed to transmit electricity, primarily generated by a gas engine. Three 170 kW gas engines and a 100 kW PV system were installed at a sewage plant. A 50 kW inverter (Figure 6.50) for the PV system compensated for demand imbalances between the three phases. Because thermal energy was in short supply and it was necessary to safeguard the microorganisms that produced digestion gas in the sewage plant, a wood waste steam boiler was also installed.

Between the sewage plant and city office, four schools and a water supply authority office were connected to the private distribution line. At the school sites, renewable energy resources were used to create a power supply that fluctuates according to weather conditions in order to verify the microgrid control system's capabilities to match supply and demand. Operation of the Kyotango Project system commenced in the middle of FY2005 and concluded in March 2008. After the finish of the project, gas engine generators were used as private generators by Hachinohe city and generated energy was sent to the electricity to city office using a private distribution line.

In the Hachinohe project, the control system used to balance supply and demand had three aspects: weekly supply and demand planning, economic dispatch control once every three minutes and second-by-second power flow control at interconnection points. The control target was a power imbalance of less than 3% for moving average six-minute intervals. From

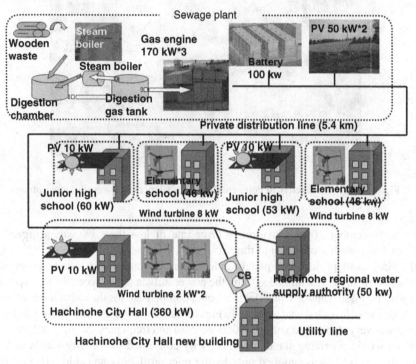

Figure 6.49 Schematic diagram of the Hachinohe project

Figure 6.50 The Hachinohe project

Figure 6.51 PV inverter to compensate for imbalances between the three phases

October to November 2006, a margin of error rate of less than 3% was achieved during
99.99% of the system's operational time.

In the Hachinohe project, completely independent operation was tested for one week in
November 2007. In the Hachinohe system, the power imbalance between the three phases was
a serious problem, so a new PV inverter that could compensate for the imbalances between the
three phases was installed and operated (Figure 6.51). Power quality during independent
operation was very stable, and voltage (6600 V ±5%) and frequency (50 Hz ±0.3 Hz) could be
maintained within operating standards. Electricity consumed in the City Hall's main office
building (Figure 6.52) was supplied only by the microgrid. No one could tell the difference
between electricity from the microgrid and electricity from the utility.

Figure 6.52 City Hall lights illuminated with power supplied solely by the microgrid

6.4.2 Network Systems Technology Projects

In the Demonstrative Project on New Power Network Systems (FY2004–FY2007), network technologies for future distribution systems were developed. This project included two experimental sub-projects (and one research sub-project). The first experimental sub-project was the "Demonstrative Project on Power Network Technology." In this project, equipment that can control the voltage and power flow to distribution feeders was developed and improved.

The second sub-project was the "Demonstrative Project on Power Supply Systems by Service Level." For this project, a test facility was constructed in Sendai City. Through this test facility, electricity of various quality levels was supplied to consumers to meet different power quality requirements. This facility was usually considered as a kind of microgrid. Construction of the project's test facility has been completed and demonstrative operations were commenced in November 2006 and concluded in March 2008.

6.4.2.1 Demonstrative Project on Power Supply Systems by Service Level

Grid technology has the potential to create value for consumers by supplying appropriate energy service levels. This is the subject being evaluated in the Demonstrative Project on Power Supply Systems by Service Level. In this project, the electricity supply system shown in Figure 6.53 was constructed in Sendai City. In this system, two 350 kW gas engine generators, one 250 kW MCFC and various types of compensating equipment were installed. The compensating equipment included an integrated power quality backup system that supplied high-quality power to "A" class consumers and "B1" class consumers, for whom interruptions and voltage drops were compensated by a UPS backup system. For "A" consumers, the wave pattern was guaranteed, whereas for "B1" consumers it was not. In the case of "B2" and "B3" class consumers, only short-term voltage drops were compensated for by a series compensator.

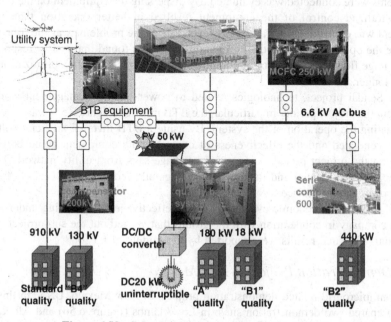

Figure 6.53 Schematic diagram of the Sendai project

Figure 6.54 Sendai project's compensating equipment and generators

The power supply system's compensators and distribution generators are shown in Figure 6.54. The equipment was tested using dummy loads in FY2006. In FY2007, the equipment was connected to meet actual demand, such as for a university or the Sendai City facility, and demonstrative operation was commenced. Moreover, back-to-back (BTB) equipment was added to the system and then applied to create an artificial voltage drop for testing the function of the compensating equipment.

In the Demonstrative Project on Power Network Technology, technologies evaluated in this project were thought to contribute to better voltage management. Through digital simulations and experiments using model distribution networks, the effectiveness of those technologies to control voltages on distribution lines when many distributed generators like PV systems were connected was evaluated. By dispersing the equipment on the distribution line, centralized control of the equipment resulted in better operation than when the equipment was controlled separately. However, one of the problems was the time it took to calculate the optimal centralized control solution. It was found that a slow response could cause voltage fluctuations if a lot of equipment was installed and the optimization interval became longer.

In the Sendai project, technologies related to power quality management worked well throughout the project period. In particular, the BTB inverter (Figure 6.55) was very effective for evaluating the operation of the system. By using a BTB inverter, artificial voltage sags could be generated and the effectiveness of compensating equipment could be evaluated. Throughout the project period, some natural voltage sags from utility networks were also experienced. The artificial and natural voltage sags allowed the effectiveness of the high-quality power supply service to be demonstrated.

In this project, an economic evaluation of cost effectiveness were being undertaken by a research company in cooperation with the groups that carried out the sub-projects in Akagi and Sendai. The final results were reported by the middle of FY2008.

6.4.3 Demonstration Project in New Mexico

NEDO ran microgrid-related demonstration projects in New Mexico, USA. In this project, NEDO prepared two demonstration sites: in Los Alamos (Figure 6.56) and Albuquerque.

Figure 6.55   BTB inverter used in Sendai project

6.4.3.1   Los Alamos County Project

In Los Alamos county, there is the famous Los Alamos National Laboratory (LANL), which is situated on the plateau known as the mesa at an altitude of almost 2200 m, and it is within a residential area of 20 000 people. It has a dry climate, but it has also a lot of thunder storms in the summertime. Electricity is supplied by a county-owned electricity utility, and its distribution network feeds the residential area by two underground cables which are connected to some feeders using overhead lines. By changing connections using switchgear, the demand on each underground cable can be changed.

Figure 6.56   View of Los Alamos County

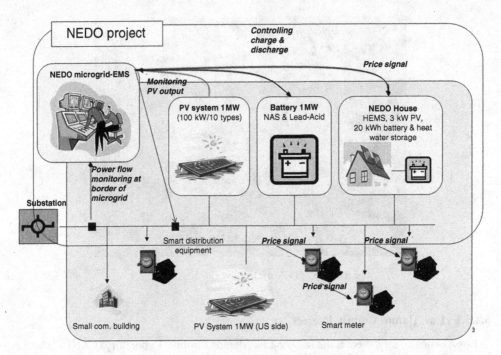

Figure 6.57 Microgrid sytem in Los Alamos

On one of those feeders, as shown in Figure 6.57, NEDO has installed a 1 MW PV system and the US authorities have installed a second 1 MW PV system. Also, Japan installed a battery storage system of over 1 MW. Using this combination, NEDO will demonstrate reduction of power fluctuation due to intermittent PV production.

At the Los Alamos demonstration site, NEDO plans to test demand response by introducing a special time of use (TOU) menu. In this case, the energy management system sends a price signal to each smart meter, in order to measure load elasticity to price. Also, using this price signal, NEDO will demonstrate a high specification home energy management system (HEMS) in the demonstration house. This HEMS involves electricity price and PV generation forecasting, based on weather forecasts. This HEMS also calculates the optimal scheduling of the battery storage operation. NEDO aimed to demonstrate that such an intelligent HEMS would bring the largest demand response.

6.4.3.2 Albuquerque Project

Albuquerque is the largest city in New Mexico. Its population is about 750 000, and the altitude is almost a mile above sea level. The sun shines for about 300 days a year, and there are thunderstorms on many summer evenings.

In Albuquerque, the Public Service of New Mexico (PNM) supplies electricity. In the southern area around the Albuquerque Airport, there is a new development area named Mesa Del Sol, where the second demonstration area is located. In the central area of Mesa Del Sol, there is a new town center building (Figure 6.58), which is the main site for NEDO's demonstration.

Figure 6.58 Town center building

By introducing a gas engine co-generation system, a fuel cell, a battery storage system and thermal storage (using ice), this building can supply ancillary services to the grid distribution company. On the grid, there is a 500 MW PV system (Figure 6.59). Several distributed generators and storage elements work as absorbing equipment to accommodate the fluctuation of the PV by receiving a signal from the EMS. Also, this building can operate independently from the grid. Throughout the project, NEDO has demonstrated that independent operation can provide a high reliability supply for such a building.

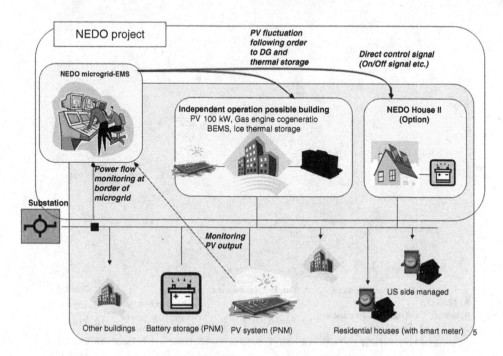

Figure 6.59 Demonstration system in Albuquerque

6.5   Overview of Microgrid Projects in China

Meiqin Mao (Section 6.5)

By the end of 2012, 16 microgrids had been installed in various parts of China, as shown in Figure 6.60. These microgrids may be divided into three groups according to their installation sites:

- 4 microgrids installed on islands along the east and south coast of China
- 8 microgrids installed at industrial, commercial or residential sites
- 4 microgrids installed in remote areas

6.5.1 Microgrids on Islands

Islands have complex energy supply problems due to their unique geographical locations and natural environment. At present, most islands in China are supplied by diesel generators, with adverse effects on supply reliability and environmental pollution, including in surrounding waters. The use of submarine cables is mostly not economical. Islands have abundant renewable energy resources, mainly solar, wind and biomass, making them ideal sites to employ microgrid technologies.

The operating microgrid projects on islands include the MW-level multi-energy microgrid system at Dong'ao island and the wind/PV/diesel and seawater desalination comprehensive

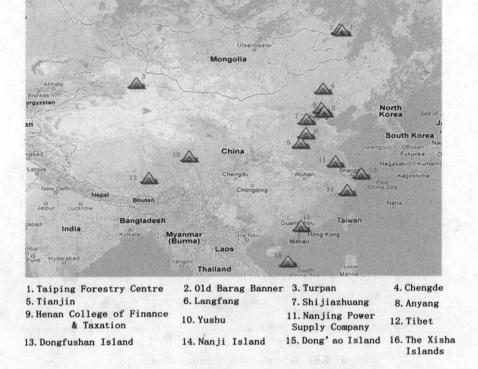

1. Taiping Forestry Centre	2. Old Barag Banner	3. Turpan	4. Chengde
5. Tianjin	6. Langfang	7. Shijiazhuang	8. Anyang
9. Henan College of Finance & Taxation	10. Yushu	11. Nanjing Power Supply Company	12. Tibet
13. Dongfushan Island	14. Nanji Island	15. Dong'ao Island	16. The Xisha Islands

Figure 6.60   Location of microgrids installed in China

demonstration microgrids at Dongfushan island. Also, a number of microgrids on islands are under construction or in planning, such as the stand-alone microgrid project at Nanji island, the new energy demonstration microgrid at Wanshan islands, Zhuhai, BYD microgrids, the grid-connected microgrid demonstration project on Luxi island and the 500 kW ocean energy stand-alone in Qingdao. These microgrids can be divided into islanded and grid-connected systems based on their operation modes. The microgrids at Dong'ao island and Luxi island are typical grid-connected systems, while the microgrids at Nanji island and Dongfushan are off-grid systems. In the following sections selected microgrids in operation are briefly described.

6.5.1.1 MW-Scale Microgrid Installed at Dong'ao Island

Dong'ao Island is located at the center of the Wanshan Islands, Zhuhai City, with an area of 4.62 km^2 and a population of over 600 [1]. Despite the abundant solar energy, wind energy, biomass energy and other renewable energy resources, the power supply used to be a bottleneck to the island's economic development and ecological preservation. In 2010, the first MW-scale microgrid with multi-energy technologies was built (Figure 6.61) and put into successful operation by the Guangzhou Institute of Energy, Chinese Academy of Science and China (Zhuhai) Xing-ye Solar Technology Holdings Co. Ltd. Its configuration includes a 1000 kW solar PV, a 50 KW wind generation system, a 1220 kW diesel power generator and a 2000 kVAh battery bank [2,3]. In addition, a control unit with a bidirectional inverter and a monitoring unit equipped with GPRS wireless communication devices is used to manage the microgrid operation. In the future, the microgrid may be extended by adding various other energy resources, such as tide and wave energy generation systems.

6.5.1.2 Stand-Alone Microgrid at Nanji Island

The stand-alone microgrid at Nanji Island, in Zhejiang, is another microgrid demonstration project. Building started in June 2012, and it is planned to be put into operation by the end of 2013. The State Grid Corporation of China supports the project with a total investment of ¥150 million. The Nanji microgrid consists of 10 sets of 100 kW wind generators, 545 kW of PV, 30 kW of marine power generation, 300 kW of storage batteries and a 1600 kW standby diesel generator. The project will also replace all internal combustion engine vehicles on the island with electric vehicles (EV). The battery system also takes advantage of the EVs' batteries which will use its stored energy to supply electricity to the system reducing the operating time of the diesel generator, making the best use of excess renewable energy [4–6]. There are about 2400 people living on the island.

Figure 6.61 DER at Dong'ao Island's microgrid

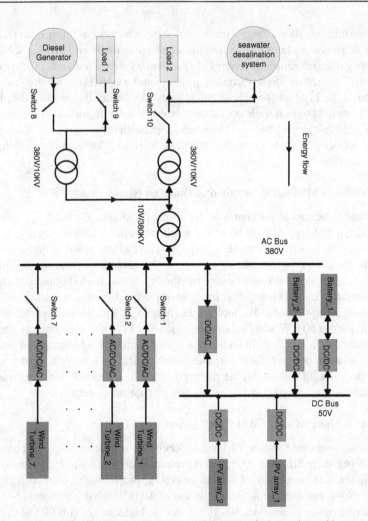

Figure 6.62 Schematic diagram of the Dongfushan microgrid

6.5.1.3 Microgrid at Dongfushan Island

The total installed capacity of the microgrid at Zhejiang, Eastern Fushan island is 300 kW. This is a hybrid wind/PV/diesel and seawater desalination system, which includes seven 30 kW wind turbines, a 100 kWp PV generation system and a set of 50 t/d seawater desalination systems (Figures 6.62–6.64). The system is also equipped with batteries. The microgrid uses hybrid AC and DC buses [4]. The operating strategy aims to maximize the use of renewable energy and by reducing the diesel power and by extending the battery service life, taking into account their characteristics. At present, it is the largest off-grid operating microgrid system in China [2].

6.5.2 Microgrids in Industrial, Commercial and Residential Areas

These projects have been installed in different places in China, typically in intensely industrial, prosperously commercial and densely populated areas, such as industrial parks,

Figure 6.63 Wind power generation system [13]

(a)   (b)

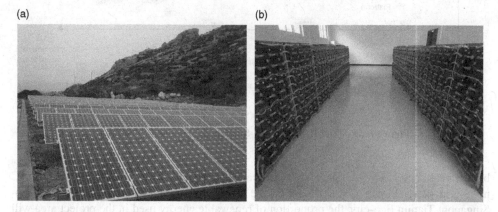

Figure 6.64 The 100 KWp PV generation system and energy storage system at Dongfushan Island [13]

university campuses, commercial or residential buildings. Their purpose is to provide safe, reliable and stable power supply to the loads, thus improving the power quality. Most of these projects are connected at 10 kV MV distribution networks and their capacity varies from hundreds of kW to 10 MW.

Among these projects is the Sino-Singapore Tianjin Eco-city smart grid demonstration project, which is China's first smart grid demonstration project, and the microgrid system in Nanjing Power Supply Company which can achieve smooth transition between grid-connected and islanded mode of operation, plus a typical village microgrid demonstration project that has been developed in Yudaokou, Chengde. In addition to these projects, microgrids in industrial, commercial and residential areas include Xin Ao (Langfang) future energy Eco-city microgrid demonstration project, the 2 MW hydroenergy/PV complementary microgrid demonstration project in Batang, Yushu, and the Anyang microgrid demonstration project.

6.5.2.1 Sino-Singapore Tianjin Eco-City Smart Grid Demonstration Project

The Sino-Singapore Tianjin Eco-city smart grid demonstration project comprises distributed energy resources of 6 kW of wind power and 30 kW of PVs, energy storage of four 15 kW

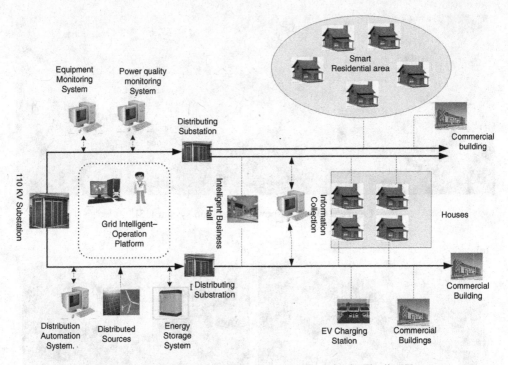

Figure 6.65 Microgrids at Sino-Singapore Eco-city in Tianjin [7]

battery banks and 15 kW lighting and electric car charging pile loads to form a low-voltage AC microgrid [2,4] (Figure 6.65). Based on the overall planning requirements of the Sino-Singapore Tianjin Eco-city, the proportion of renewable energy used in the project area will exceed 20%. After its completion, the total installed capacity of the microgrid will reach 175 MW, and it is expected that the annual generation capacity will be about 390 million kWh to meet the electricity demand of about 130 000 families [7].

6.5.2.2 Wind/PV Microgrid System at Nanjing Power Supply Company [2,4]

The wind/PV microgrid designed by Zhejiang Electric Power Test Research Institute in the Nanjing Power Supply Company, is one of the first microgrid projects to be put into operation in China. The goal of the project is to study the coordinated and optimal control technologies when several DER (PV, wind power and batteries) are connected to the local grid. The microgrid includes a 50 kW solar PV, 15 kW wind generation (two wind turbines of 8 kW and 7 kW, respectively) and a 300 Ah battery bank (Figures 6.66 and 6.67). Using intelligent control of the charging/discharging process of the battery systems, this system can smooth out the distributed generation (DG) power output variations caused by fluctuating solar radiation and wind speed and is able to operate in both, grid-connected and islanded modes. In grid-connected mode, the system can control the power flow at point of common coupling (PCC) by controlling the charging/discharging of batteries, while in islanded mode, the battery banks supply the main power. Static switches are used for the seamless transition between the two modes.

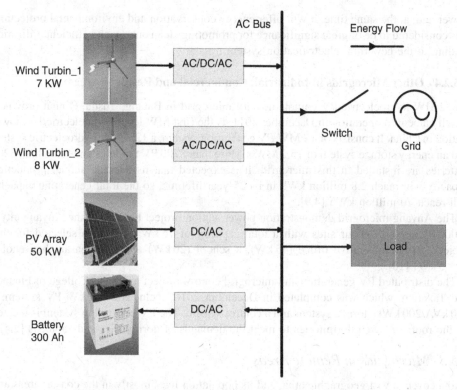

Figure 6.66 Microgrid at the Nanjing Power Supply Company

Figure 6.67 Wind/PV microgrid system at the Nanjing Power Supply Company [13]

6.5.2.3 Microgrid in Yudaokou, Chengde [8]

This is a village microgrid of 110 kW capacity. It contains 60 kW wind power generation, 50 kW PV and 80 kW (128 kWh) storage and comprises four sets of household wind/PV/battery microgrid. After completion, it will be able to provide local farmers with reliable power and, at the same time, it will offer energy conservation and environmental protection. It is considered to be of great significance for promoting clean energy and efficient utilization leading to the new rural electrification system.

6.5.2.4 Other Microgrids in Industrial, Commercial and Resident Areas

The 2 MW hydroelectric/PV complementary microgrid in Batang, Yushu, Qinhai province, which came into operation in December 2011, is the first MW level hydroelectric/PV power station in China. It consists of a 2 MW PV generation system, a 12.8 MW hydroelectric system and an energy storage system of 15.2 MWh. More than 8700 PV modules and more than 8200 batteries are installed in this microgrid. It is expected that its average annual production capacity will reach 2.8 million kWh in its 25-year lifetime, so the total generating capacity will reach 70 million kWh [4,9].

The Anyang microgrid demonstration power station project belongs to the Anyang city's "3000 PV project". Four sites with a total capacity of 75 kW have been selected for this project, an administrative office (45 kW), a school (20 kW) and two peasant households (2 × 5 kW) [10].

The distributed PV generation and microgrid control project in Henan College of Finance and Taxation, which was completed in December 2010, includes a 350 kW PV system, a 200 kVA/200 kWh storage system and a control system. The PV generation system is located on the roofs of seven dormitories to meet the demands of dormitories and canteens [2,4].

6.5.3 Microgrids in Remote Areas

China covers a vast geographic area, and its population lives mostly in the coastal areas and middle or eastern areas inland. Some areas, such as Tibet, Qinghai, Sichuan, Yunnan, Xinjiang, Inner Mongolia and other western and frontier places are off the power grid, with 1.2 million households and 5 million people without electricity. Grid expansion to these areas faces several difficulties, such as huge construction investments and high operational costs. The installation of microgrid projects based on local renewable energy sources, such as solar and wind power, will not only make full use of the unique natural advantages and improve energy efficiency, but can also provide local residents with an electrified way of life. Such remote area microgrids are the New Energy City Demonstration Project in Turpan, a microgrid with wind/PV/battery in Old Barag Banner and Taiping Forestry Center, and a microgrid with PV generation in Ali area of Tibet.

6.5.3.1 Microgrid of New Energy City Demonstration Project in Turpan [2,4]

Turpan is an ideal place for the development of renewable energy in China because of its unique natural conditions. It is located in Xinjiang province with abundant solar and wind resources. The first construction phase of a project called "Microgrid of New Energy City Demonstration Project in Turpan" has been completed in 2012 [11]. Partners include the National Energy Administration, the National Development and Reform Committee and the State Electricity Regulatory Commission. The project consists of an integrated PV project (a rooftop PV power station), which uses the 750 000 m^2 rooftop to install 13.4 MW PV array, forming the Building Integrated Photovoltaic Project. The project consists of 57 000 poly-silicon 235 W PV modules, 735 small-sized inverters, a monitoring system and related accessories. The smart microgrid project includes a 10 kV substation, 380 V power distribution network, 1 MWh storage system, electric bus charging station, monitoring control center for the microgrid, ancillary works and so on.

6.5.3.2 Other Microgrids in Remote Areas

The microgrid at Old Barag Banner is a typical grid-connected microgrid. The system includes 150 kWp PV power generation, 100 kW wind power generation, and a 50 kW × 2 h lithium-ion battery system. The microgrid at Taiping Forest Center is an off-grid microgrid with 100 kW wind power, 250 kWp PV and a 200 kW × 4 h battery system [2]. The PV microgrid at Ali has a total installed capacity of 10 MW and its design life is 25 years. It is equipped with presently the largest energy storage system in the world of 10.64 MWh. Together with an existing 6.4 MW hydropower and 10 MW diesel generator, they form a stand-alone PV/hydropower/diesel microgrid [12].

For further information, see [14–17].

References

1. http://www.zhnews.net/html/20110516/073516,295032.html. "The microgrid at Dongao Island have product 120 000 kWH electricity".
2. Zhang, Jiajun and Zhao Dongmei (Sep. 2012) "Overview of Micro-grid experiment and demonstration projects at home and abroad". 28th academic annual conference.
3. http://www.gd.chinanews.com/2010/2010-12-28/2/79074.shtml. "China'First MG level intelligent microgrid based on island built in Zhuhai".
4. http://news.byf.com/html/20121212/154422.shtml. "Overview of Microgrids in China".
5. http://www.stdaily.com/kjrb/content/2012-06/06/content_477760.htm. "National 863 Plan" Project–Microgrid in Nanji island is put into construction.
6. http://www.sgcc.com.cn/ztzl/newzndw/sdsf/06/274841.shtml. "The effect of microgrid is not micro—Explore the future dream of Nanji Island."
7. http://www.tianjinwe.com/tianjin/tjwy/201110/t20111026_4466096.html. "Explore The Sino-Singapore Tianjin Eco-city comprehensive smart grid demonstration project".
8. http://www.ceeia.com/News_View.aspx?newsid=39467&classid=81. "Microgird provides a new way for integrating new energy into grid".
9. http://www.stdaily.com/kjrb/content/2012-01/06/content_410839.htm. "The Hydroelectric/PV complementary power station in Yushu is put into use successfully".
10. http://zs.cctv.com/anyang/news_209474.html. "The 45kW Anyang microgrid demonstration power station is put into grid-connected operation successfully".
11. http://www.xjbs.com.cn/news/2012-11/10/cms1484505article.shtml?nodes=_365_.
12. http://www.windosi.com/news/414733.html. "World's highest 10 MW-level photovoltaic power station is put into operation".
13. http://wenku.baidu.com/view/99cbfcd95022aaea998f0f85.html.
14. http://www.sgcc.com.cn/xwzx/gsyw/yxfc/11/284269.shtml, "The first mrciogird of Shandong will be built in Qindao".
15. http://news.sina.com.cn/o/2012-08-28/152225052783.shtml. "China Southern Power Grid will built microgrid project in Xisha Islands".
16. http://www.wenzhou.gov.cn/art/2012/12/26/art_3907_249918.html. "The system of Luxi microgrid is being built".
17. http://www.zhnews.net/html/20110928/081327,324712.html.

6.6 An Off-Grid Microgrid in Chile

Rodrigo Palma Behnke and Guillermo Jiménez-Estévez (Section 6.6)

6.6.1 Project Description

The ESUSCON rural electrification project (Electrificación Sustentable Cóndor, or Condor Sustainable Electrification Project, in English) is an initiative from the Energy Center, Faculty of Physical and Mathematical Sciences, Universidad de Chile, and a local mining company, focused on developing a microgrid in the locality of Huatacondo, Tarapacá region, Chile.

The Huatacondo's electricity network is isolated from the interconnected system. Electricity supply was limited to only 10 hours a day using a diesel generator. The developed microgrid takes advantage of the area's distributed renewable resources, providing 24-hour electricity service. Because the village experienced problems with its water supply system, a management solution was also included in the microgrid. Also, a demand-side option to compensate the generation fluctuations due to the renewable sources was considered. The system is composed of two PV systems (Figure 6.68), a wind turbine, the existing village diesel generator unit (typical in isolated locations), an energy storage system (ESS) composed of a lead-acid battery bank (LABB) connected to the grid through a bidirectional inverter, a water pump and a demand side management (DSM) (loads). Figure 6.69 shows the power schematic of this microgrid.

These elements are controlled by a central energy management system (EMS) that provides signals for optimizing their operation according to load and resource forecasts, in order to minimize the consumption of diesel and to keep the power quality indicators close to optimal values. Also, it includes a demand-side management system, which sends to the customer visual information about recommended daily load profiles according to forecast resource availability; actual consumption data is then recorded and sent back to the EMS through smart metering [2].

The main goals of the EMS are

- minimize the use of diesel
- deliver active generation setpoints for the diesel generator, the ESS inverter and the PV plant

Figure 6.68 The Huatacondo microgrid

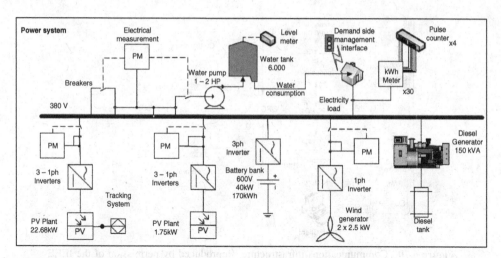

Figure 6.69   Renewable-based microgrid: Power schematic. Reproduced by permission of the IEEE

- turn on and off the water pump in order to keep the elevated water tank level within predefined limits
- send signals to consumers promoting behavior changes.

In field implementation, the diesel generator and the ESS inverter have two configurable droop curves to follow the setpoints. Q–V and P–f droop curves are normally configured for typical operation states that change when receiving adequate signals from the EMS.

When the EMS turns on the diesel generator, it is convenient to start the LABB charging. In that case, the diesel generator is configured to work in synchronous mode (with infinite slope in both curves, takes all the variations between generation and load, following the rated values of 380 V_{1-1} at 50 Hz, and the ESS inverter curves are configured to follow the charging profile of the batteries.

When the diesel generator is turned off, the droop curves of the ESS inverter change to those for a master operation mode, with a very low droop characteristic.

6.6.1.1   Technological Features

The communications infrastructure is devoted to acquiring field data for optimizing the operation of the system. Thus, a SCADA and several measurement devices were installed at Huatacondo village (Figure 6.70). This SCADA has the following capabilities:

- electrical variable measurements for all generation units
- some electrical measurements in the grid and control capabilities (connection/ disconnection of low voltage network feeders)
- energy consumption measurement of the electrical loads in the network
- power control of the energy storage system inverter and diesel generator
- grid connection control for all generation units and the water pump consumption

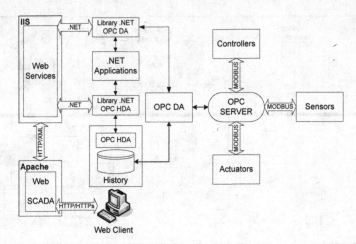

Figure 6.70 Communications infrastructure. Reproduced by permission of the IEEE

- sun tracking control for the principal PV plant
- wireless communication with the interfaces of the demand-side management system

6.6.2 Demand-Side Management

Online signals are sent to consumers in order to modify their consumption behaviours, leaving daily energy constant. This modification is modeled by shifting coefficients $S_L(t)$ that are provided by the EMS.

The application of the DSM, relies on an interface that consists of a 24 hour clock that assigns a specific color, describing the energy availability, for each hour. For instance, red

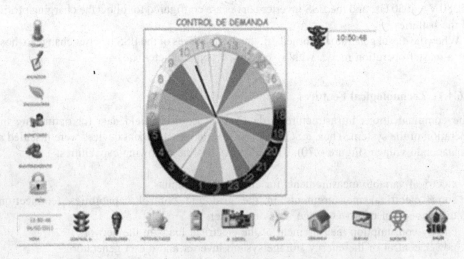

Figure 6.71 DSM interface. Reproduced by permission of the IEEE

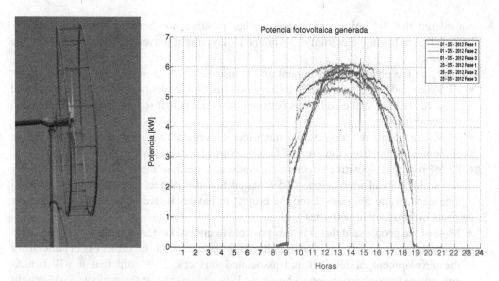

Figure 6.72 Bird protective device (left), comparison between PV panel without (continuous line) and with solar tracking system (dotted line)

means basic energy consumption, green complete energy availability and yellow is a warning state. The DSM interface is shown in Figure 6.71.

6.6.2.1 Other Developments

- *Bird protective device*: In Huatacondo there is the continuous presence of the Andean condor, *Vultur gryphus*. In order to avoid the possible collision of this bird with the wind power turbine it was necessary to design a special protective cage, which is assembled onto the turbine nacelle. This protective device looks very similar to those used for common fan applications ensuring (Figure 6.72 left).
- *PV sensorless automatic tracking system for sun model improvements*: a single-axis tracking system was developed for this microgrid, the main advantage being that this system avoids local control, and instead setpoints are defined by the microgrid EMS. In Huatacondo, given the high radiation values, the PV plant capacity factor may increase to 5%, with the tracking system implementation (Figure 6.72 right).

6.6.2.2 Identified Positive Impacts

A survey of perceptions of the impact of the Huatacondo project at the community level was conducted, covering 95% of the households in August 2011, almost a year after commissioning was completed.

The results of the survey indicate that the microgrid has high acceptance within the community.

1. Impacts on the community
   The project has a positive effect on daily life according to 73% of the people, since they can now perform more recreational and economic activities that require electricity. Just 20%

considered that the microgrid brings neither positive nor negative effects, while 7% believed that having electricity 24 hours a day might have negative effects on the community.

2. Impacts on the physical environment and landscape
   - 57.5% considered the microgrid to have no effect on the fauna, while 42.5% expressed concern about possible collisions between the condors that regularly fly in the area and the wind turbine.
   - 92.5% stated that the project does not affect the vegetation, since the installation of the equipment is far from the area of farms, crops and native vegetation.
   - 82.5% reported a positive effect on the landscape, because the project brings modernity and technological innovation, 12.5% stated that the project does not affect the local landscape, while 5% considered the project to have negative effects on the landscape.

3. Impacts on productive activities
   - Most of the people said that it is now possible to undertake new economic activities or to enhance the existing ones. Specifically, 27% believed that the project is beneficial for the development of tourism and associated services; 23% said that it will benefit agriculture through irrigation technology; 18% suggested that the project would benefit construction activities now that the daily period in which they can use electric tools has been extended. Finally, the remaining 32% felt that there could be a negative impact on agricultural development, because farmers might neglect their farms and crops while pursuing other activities that are now possible.

   Over the time that the microgrid has been operating, the operators have gathered the following evidence:
   - no equipment failures over the 2010–2011 period
   - high level of commitment of the community for maintenance activities and demand response programs
   - high level of interest shown by the people in charge of maintenance and operation activities, demonstrated by low rotation in the membership of the maintenance team
   - no accidents involving people or animals

   Additionally, the system provides other operational benefits:
   - diesel consumption reduced by 50% when compared with the previous operation scheme
   - improved reliability levels
   - higher power quality

References

1. Alvial-Palavicino, C., Garrido-Echeverría, N., Jiménez-Estévez, G., Reyes, L., and Palma-Behnke, L. (2011) A methodology for community engagement in the introduction of renewable based smart microgrid. *Energy for Sustainable Development*, **15** (3), (SI), 314–323.
2. Palma-Behnke, R., Reyes-Chamorro, L., and Jimenez-Estevez, G. (2012) Smart solutions for isolated locations. IEEE PES General Meeting.
3. http://synergy.ucsd.edu/files/Agarwal_DATE2011_UnderstandingMicroGrid.pdf.

7

Quantification of Technical, Economic, Environmental and Social Benefits of Microgrid Operation

Christine Schwaegerl and Liang Tao

7.1 Introduction and Overview of Potential Microgrid Benefits

Depending on its operation strategy (see Figure 1.5), a microgrid can provide a large variety of economic, technical, environmental, and social benefits to both internal and external stakeholders. This chapter analyzes such benefits in detail, based on typical networks identified for a number of representative EU member states. The study has been completed in the framework of the EU R&D project, More Microgrids [1].

Figure 7.1 provides an overview of microgrid benefits in economic, technical and environmental aspects. Each benefit item is mapped to the related stakeholder with dotted lines. Obviously, identification of microgrid benefits is a multi-objective and multi-stakeholder interest coordination task.

Before going into details of a quantitative analysis, a brief qualitative overview of these benefits is given.

7.1.1 Overview of Economic Benefits of a Microgrid

The economic values created by a microgrid can be roughly categorized into providing **locality** benefit and **selectivity** benefit. Locality benefit is mainly attributed to the creation

Microgrids: Architectures and Control, First Edition. Edited by Nikos Hatziargyriou.
© 2014 John Wiley & Sons, Ltd. Published 2014 by John Wiley & Sons, Ltd.
Companion Website: www.wiley.com/go/hatziargyriou_microgrids

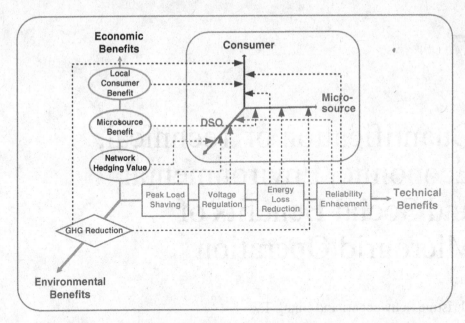

Figure 7.1 Overview of microgrid benefits

of an internal "over-the-grid" energy market within the microgrid, where microsource units could sell at prices higher than the prices at wholesale level and end consumers could buy at prices lower than the retail level. This is equivalent to Market Policy 1 (good citizen behavior), analytically described in Section 2.7.2. Selectivity benefit is associated with optimization of real-time dispatch decisions that minimize opportunity cost of the whole microgrid with consideration of technical and environmental constraints. This is equivalent to Market Policy 2 (ideal citizen behavior), analytically described in Section 2.7.2.

Both locality and selectivity benefits can be attributed to either the consumer side or the microsource side, where a proper market design/interest allocation mechanism is needed to ensure a reasonable split of both benefits between consumer and microsource owner. From a macro-economics point of view, the locality benefits of a microgrid can be grouped into two aspects:

- microgrid can act as an initiator of local retail and service markets,
- microgrid can act as a hedging tool against potential risks of price volatility, outage, load growth and so on.

The selectivity benefit of a microgrid can also be further categorized in the following two aspects:

- microgrid can act as an aggregator of both supply- and demand-side players,
- microgrid can serve as an interest arbitrator for different stakeholders.

7.1.2 Overview of Technical Benefits of a Microgrid

A microgrid can potentially improve the technical performance of the local distribution grid mainly in the following aspects:

- energy loss reduction due to decreased line power flows,
- improved voltage quality via coordinated reactive power control and constrained active power dispatch,
- relief of congested networks and devices, for example during peak loading through selective scheduling of microsource outputs,
- enhancement of supply reliability via partial or complete islanding during loss of main grid. When the total number of microgrids reaches a sufficiently high share in LV substations, similar technical benefits can be expected in upstream grids as a consequence of multi-microgrid operation.

The actual level of technical benefits depends mainly on two factors: the optimality of microsource allocation and the degree of coordination among different players. Effective planning of microsource size and location can maximize unit contribution to system performance and random interconnection of oversized microsources in weak grids, but it can create more technical problems than the benefits it provides. It is also critically important that a real-time, multi-unit coordination platform, in either centralized or decentralized form (see Chapter 2) is needed to maintain targeted microgrid technical performance at all times.

7.1.3 Overview of Environmental and Social Benefits of a Microgrid

Environmental benefits of a microgrid can be expected from two aspects: shift toward renewable or low-emission (e.g. natural gas) fuels and adoption of more energy-efficient energy supply solutions (e.g. combined heat and power applications) including demand-side integration. With widespread national support policies for distributed renewable resources – for example, photovoltaics – the fuel-switching credit of microgrid is expected to grow as RES costs decrease over the years. Application of CHP and district heating and/or cooling concepts, on the other hand, varies significantly from region to region and is expected to find considerably different levels of acceptance across Europe. Demand side response, either manual or automated, is also greatly facilitated by the controlled operation of a microgrid.

Social benefits of microgrids can be mainly expected from:

- raising public awareness and fostering incentives for energy saving and GHG emission reduction,
- creation of new research and job opportunities,
- electrification of remote or underdeveloped areas.

All of these listed impacts, however, can be seen as long-term effects, and although qualitatively easy to understand, can be very difficult to quantify.

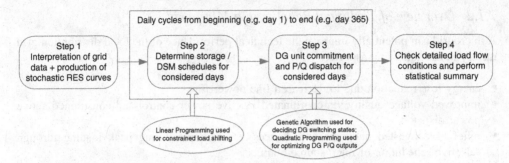

Figure 7.2 Algorithm overview of microgrid benefit quantification study

7.2 Setup of Benefit Quantification Study

7.2.1 Methodology for Simulation and Analysis

For this book, an analysis of different microgrid benefits was carried out on a yearly basis using the sequential Monte Carlo simulation method (Figure 7.2). Due to the cyclic nature of both electricity market prices and renewable (e.g. photovoltaic) output, intraday microgrid schedules are created and evaluated as hourly settlement results. Annual outcome of the optimization algorithm is consequently a statistical summary of individual daily simulations. In this section, the general framework of this analysis is outlined.

Typical annual profiles of load, RES generation and market prices have been synthesized, based on measurements or historical data. As microgrids are typically located at LV level, with mainly small residential or commercial customers, there are significant variations in load curves, which, in combination with the uncertainties from RES output as well as (wholesale) electricity price, turn the day-ahead microgrid scheduling task into a highly stochastic problem. Since prediction errors are inevitable, potential microgrid benefits evaluated from simulated or forecast data should be viewed as best-case results – that is, actual operation decisions might be sub-optimal, thus leading to lower benefits.

In order to deal with the modeling requirements of different microsource types, RES units, storage and demand-side response resources are dispatched at priority, and then the DG unit schedules are created for active and reactive powers, respectively. For the DG unit commitment, genetic algorithms or simply priority lists can be used to determine the optimal DG on/ off states. Quadratic/linear programming models can be used for optimal power dispatch, as discussed in Section 2.7.4.

The formulation described in Section 2.7.4 can be further expanded by assigning cost functions to greenhouse gas (GHG) emissions and technical aspects, such as losses or reliability of supply and external factors can be internalized within the economic criteria, leading to an economic formulation of the optimal control strategy problem that considers:

- retail revenues collected from end customers,
- revenues of selling electricity to wholesale market,

- total cost of buying electricity from wholesale market,
- overall generation cost of DG (fuel, startup/shutdown, operation and maintenance cost),
- total emission cost from electricity generation,
- cost of energy losses in the grid,
- total cost of electricity generated from RES units,
- total control and communication costs for operating the microgrid.

Obviously, this total opportunity cost objective function is an artificial entry in a liberalized electricity market, as it represents the interests of multiple entities, and, in practice, requires extensive collaboration and compromise among the various actors.

For the optimization problem, three sets of system operational constraints are applied:

1. active and reactive energy balances in the microgrid, including the connection to the upstream network,
2. DG physical constraints, regarding minimum on and off operation time (thermal constraints) and minimum and maximum energy outputs,
3. network constraints – that is, voltage limits and overload limits of transformers and circuits.

After creation of a complete schedule, load flows are run to calculate power losses and to check voltage and loading conditions in the network. Due to the large amount of output data, network variables are recorded as discrete probability density functions via a statistical summary procedure.

7.2.2 European Study Case Microgrids

It is not possible to quantify microgrid benefits considering a "typical" average system. Constitution and performance of microgrids are generally subject to a huge variety of internal and external factors, which can be summarized under the following three categories:

1. **geographical location**, which determines types of power network, RES productivity, energy portfolios for electricity generation and electricity prices,
2. **time**, which determines microsource penetration scenarios and cost developments,
3. **sensitivities**, which relate to individual market, regulatory and operational impact factors.

To quantify microgrid benefits at European level, an evaluation framework was developed to accommodate the majority of input and output parameters at a manageable level of data complexity. Based on this framework, typical European distribution networks of different types were studied, as shown in Figure 7.3. The results of this study are presented in Section 7.3. These results are critically influenced by the assumptions listed in the following subsections of Section 7.2, and therefore should be treated only as indications of future trends and the type of studies required.

Installed generation capacity and annual energy consumption vary significantly across the different European countries examined. Figure 7.4 demonstrates the basic assumptions made in the study: average wholesale prices are 40–90 €/MWh, national CO_2 emission levels vary between 400 kg and 800 kg/MWh and end user retail electricity prices (split into energy, network charges and taxes) are 6–28 €ct/kWh.

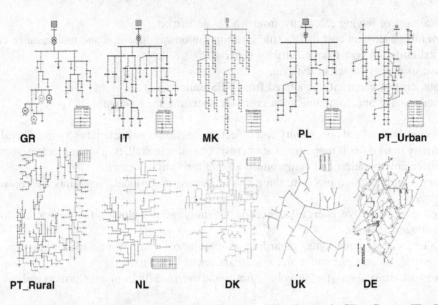

Figure 7.3 European study case networks (DE = Germany, DK = Denmark, GR = Greece, IT = Italy, MK = FYROM, NL = The Netherlands, PL = Poland, PT = Portugal, UK = United Kingdom)

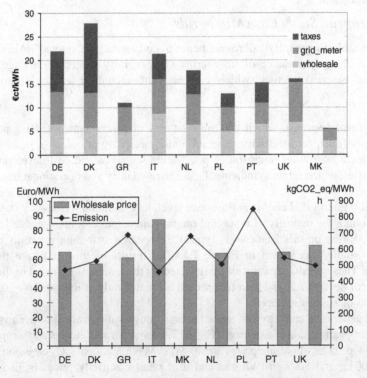

Figure 7.4 Retail/wholesale tariff structures and emission levels considered

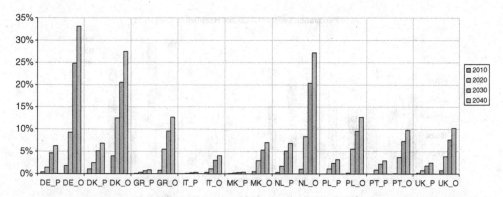

Figure 7.5 Microgrid dissemination ratio in EU national grids (O = optimistic assumptions, P = pessimistic assumptions)

7.2.2.1 Development of European Microgrids from 2010 to 2040

For the adopted evaluation framework, four future microgrid scenarios are defined for the years 2010, 2020, 2030 and 2040, that satisfy 10, 20, 30 and 40%, respectively, of their own demands by RES and bio-fueled heat-driven CHP units. This facilitates cross-national comparison, despite the regional differences in terms of adopted microsource technologies. In order to minimize the potential number of simulation cases, microsource allocation is assumed to be optimal for all cases. The RES and CHP self-supply level of the microgrids annual energy needs is kept below 50%. This is essential, since a higher level of self-supplied energy ratio would imply that the RES power installations (i.e. penetration level in terms of power rating) should be higher than 150% in countries where microgrids are dominated by intermittent RES technologies with lowfull load hours (FLH) (< 2000 h/a). The FLH is defined as the equivalent number of hours that DGs would operate under nominal capacity, in order to provide the same amount of annual energy. As active power outputs from intermittent RES and heat-driven CHP units are not naturally dispatchable, the controllability of a microgrid relies mainly on the installation of dispatchable microsources and storage devices.

After definition of a microgrid configuration aimed at satisfying local energy needs, the share of typical microgrids within national power systems (microgrid dissemination ratio) per country and region are assumed to be as shown in Figure 7.5, for 2010, 2020, 2030 and 2040, with pessimistic (P) and optimistic (O) assumptions.

Detailed penetration scenarios of RES and CHP units assumed in each simulated case are given in Figure 7.6. It is noted that countries with high RES energy production from PV (IT, PT) or wind turbines (PL) are likely to have microgrid configurations of more than 100% RES penetration levels (installed microsource capacity compared to maximum load) by 2040.

7.2.2.2 Basic Cost Assumptions

Figure 7.7 shows the range of cost/kWh microsource generation, assumed for different technologies between 2010 and 2040. High cost reductions are expected for PV and fuel cell

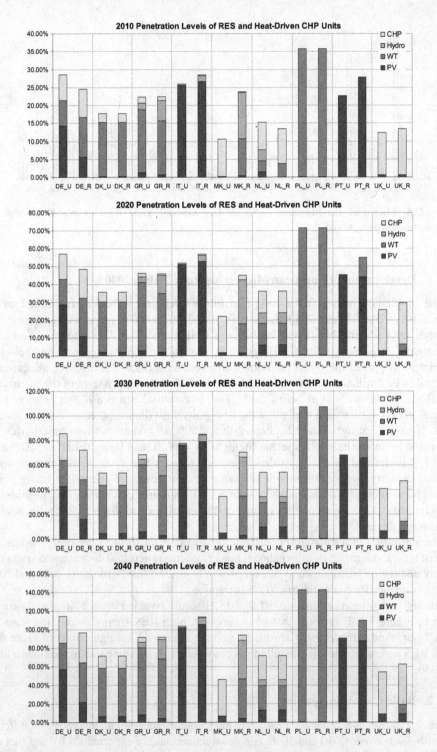

Figure 7.6 Typical microgrid configuration per country (U = urban networks, R = rural networks)

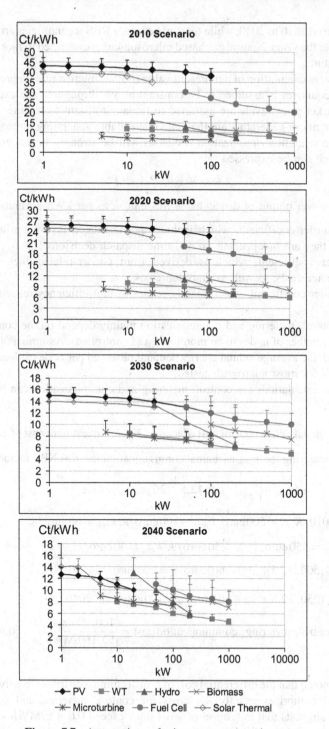

Figure 7.7 Assumptions of microsource generation costs

technologies from 2010 to 2040, while the costs of other RES technologies are assumed to be more stable over the years. Natural gas based microsources are assumed to face increasing fuel costs in the future.

For a consistent calculation of net present value (NPV), interest rate (depreciation rate) of all examined countries is assumed to be constant at 5% throughout the examined period. Electricity market price levels are assumed to remain constant over time.

Dispatchable micro-generators are the main controllable generating units within a microgrid, so system operation optimization decisions will be strongly dependent on their cost functions, which can be expressed as:

$$k = a + b \cdot P + c \cdot P^2 \tag{7.1}$$

P as active power output of dispatchable MS unit, k as per kWh generation cost

- a constant order coefficient, which includes power-related investment and installation costs as they are independent on real-time dispatch decisions
- b first-order coefficient, which is derived from energy-related fuel, operation and maintenance costs, as well as emission costs
- c second-order coefficient, which mainly relates to unit efficiency related costs.

Costs for control, metering and communication mainly depend on the complexity of the microgrid (the number of nodes to be monitored and controlled). Assuming 5000 h/a as FLH of load demand, the average annual energy demand observed per node (as load substation) is 50–160 MWh/a for most microgrids tested.

Ensuing cost calculations for control, metering and communication can be described as follows:

Define: k_{act} as annual active control cost of a node, K_{ctrl} as investment cost of control device

k_{met} as annual metering fee, k_{com} as annual communication fee, β as NPV factor

Then: $k_{act} = \dfrac{K_{ctrl}}{\beta} + (k_{met} + k_{con})$, $\beta \cong 12.5 \dfrac{1}{a}$ (20 years lifetime)

Assume: For 2010, $K_{ctrl} = 500$ euro, $k_{met} = 100$ euro/a, $k_{con} = 60$ euro/a,

For 2040, $K_{ctrl} = 250$ euro, $k_{met} = 50$ euro/a, $k_{con} = 30$ euro/a,

Then: $k_{act} = \begin{cases} (500/12.5 + 100 + 60)\text{euro}/a = 200\,\text{euro}/a \text{ in } 2010 \\ (250/12.5 + 50 + 30)\text{euro}/a = 100\,\text{euro}/a \text{ in } 2040 \end{cases}$

$\Rightarrow$ Per MWh control, metering, communication cost $= \dfrac{[100, 200]\text{euro}/a}{[50, 160]\text{MWh}/a} = [0.6, 4]\text{euro}/\text{MWh}$

$$\tag{7.2}$$

It should be noted that the differential cost of transforming a distribution network with DER to a microgrid comprises exactly the costs for control, metering and communication. Equation (7.2) suggests that microgrid benefits must exceed 0.6–4 €/MWh to justify active control measures. As this additional cost is not dependent on microsource operation decisions, it is used for comparison with the estimated microgrid benefits.

7.3   Quantification of Microgrids Benefits under Standard Test Conditions

7.3.1   Definition of Standard Test Conditions

Due to the comparatively large number of different assumptions that could impact microgrid benefits, a basic standard test condition (STC) is defined to evaluate microgrid benefits under a "most likely to happen" scenario. Mid-level wholesale market price levels corresponding to 80% of the peak in the past decade are assumed per country (Figure 7.4). Real-time and directional price setting is applied, and a combined operation mode – as defined in Section 1.5.3 – is assumed. Sensitivity analysis with regard to these preset conditions is performed in this section to evaluate microgrid performance under various evaluation indices.

7.3.2   Balancing and Energy Results

Figure 7.8 demonstrates the percentage of local load supplied by microsources and the share of microsource generation in the supply of local loads for different countries and different

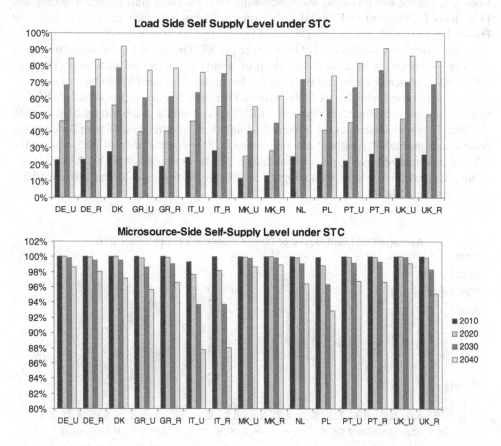

Figure 7.8   Microgrid self-supply level on demand and supply side (U = Urban networks, R = Rural networks)

microgrid setups under STC. The load-side self-supply level is the percentage of microsource-supplied load demand over the total annual load. The microsource-side self-supply level is the ratio of microsource-generated electricity that is consumed by local load as a percentage of their total generation.

Most microgrids start with a load-side self-supply level of 15–25% in 2010 and reach 70–90% of local demand by 2040. FYROM is the only case where the self-sufficiency level starts from 10% in 2010 and ends up with around 60% in 2040, mainly due to lower wholesale and retail prices, leading to low microsource utilization. Even with high microsource penetration levels (up to the case of 2040), the differences between retail and wholesale prices lead microsource units to supply local loads most of the time. A general local supply ratio of >95% can be observed with the exception of Italy, where high electricity prices tend to encourage microsource units to export more often than the average level seen in the other countries.

Figure 7.9 illustrates the full load hours (FLH) of dispatchable microsource units under STC. It is noted that countries with high initial FLH values in 2010 (>6000 h/a) generally experience a reduction of FLH as time progresses to 2040. Countries starting with low FLH value (i.e. Greece and FYROM) show increasing FLH values as time passes. The very low FLH hours (<3000 h/a) of FYROM fully explains its low self-sufficiency level shown in Figure 7.8.

Figure 7.10 shows the estimates of the average per-kWh premium support levels (on top of microgrid earnings) in order to ensure the profitability of subsidized RES units. The high initial values noticed in Italy and Portugal can be mainly explained by the high penetration levels of expensive technologies (PVs) assumed for both countries. It is shown that in the majority of countries, financial supports for RES units within a microgrid by 2030 or 2040 will largely not be needed, as the internal trading prices are already sufficient for ensuring general unit profitability without market distortion. Slower adaptation of RES units to the fully commercialized microgrid internal market can be explained either by the high costs of the assumed microsource technology (e.g. PV for Portugal) or by the low electricity prices (e.g. in FYROM).

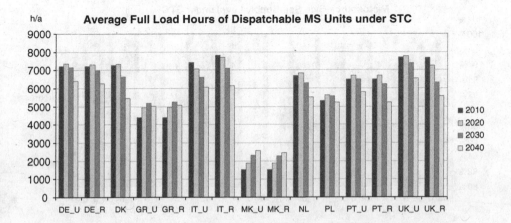

Figure 7.9 Full load hours of dispatchable microsource units under the standard test conditions

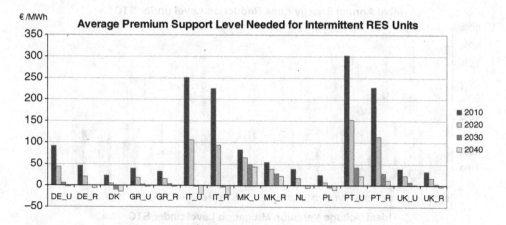

Figure 7.10 Premium support level for RES units under STC

7.3.3 Technical Benefits

7.3.3.1 Reduction of Losses and Voltage Variations

DER can provide a number of technical benefits in the operation of the distribution network, even if they are not actively controlled (passive DER operation). However, the coordination platform provided by microgrids enables the extension of these benefits to more sections of the grid, at more critical times and for serving simultaneously multiple purposes. Therefore, the technical benefits of a microgrid result from the coordinated control of all installed DER units. This probably calls for more complicated reward schemes for the DER owners.

Figure 7.11 shows the loss reduction and the voltage variation reduction in the LV network within a microgrid. Reduction of system losses (DSO perspective) refers to the reduction of annual energy losses under microgrid operation compared to the passive DER operation, that is:

Reduction of system loss (%) = 1 − (annual energy losses under microgrid)/(annual energy losses under passive grid operation)

Reduction of voltage variation refers to reduced maximum absolute voltage deviation at worst voltage quality node(s) under microgrid operation compared to the passive DER operation, that is:

Reduction of voltage variation (%) = 1 − (max absolute voltage variation in microgrid)/ (max absolute voltage variation under passive grid operation)

Figure 7.11 corresponds to the ideal situation of optimal allocation of microsources. A good correlation of loss reduction to (load-side) self-sufficiency level. Figure 7.8 shows that microgrids with optimally located microsources present significantly lower energy losses, since the majority of load demand is met locally by the microsource units. In practice, however, the DSO or microgrid operator has little or no control over the size and location of the DER. This means that the calculated loss reductions may have only a theoretical reference value. In fact, in some extreme cases, interconnection of disproportionally large microsource units at weakly loaded networks might increase energy losses. Therefore, it is very important that the microgrid planner or regulator provides the proper incentives, that could ensure efficient and effective microsource interconnection schemes to maximize the technical

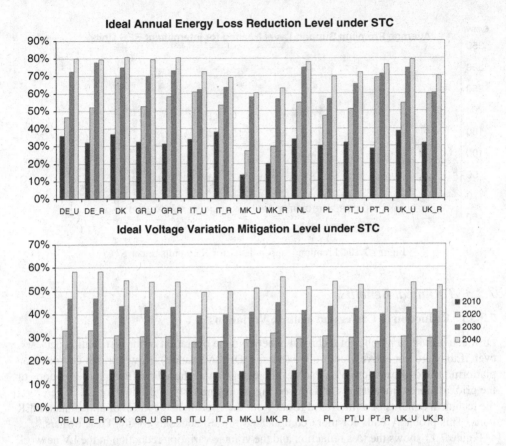

Figure 7.11 Annual energy loss reduction and voltage regulation credit

benefits from the microgrid operation. This also applies to the maximization of voltage and loading improvement benefits.

It should be noted that high R/X ratios of distribution lines result in the close coupling of active power with voltage magnitudes, as analyzed in Chapter 3. Load flow simulations of the studied networks have shown that the reactive power control of the microsource units could only contribute approximately 10–30% of the total voltage regulation. Since the active power output of the intermittent RES units is assumed non-controllable, the voltage is controlled primarily by the dispatchable microsources. This is the main reason why the effects on microgrid voltage control are less prominent.

7.3.3.2 Peak Load Reduction

Reduction of peak loading (DSO perspective) refers to the reduction of peak loading in the worst-loaded branch under microgrid operation compared to passive grid operation, that is

$$\mathrm{RPL}(\%) = 1 - (\text{peak loading in microgrid})/(\text{peak loading in passive grid operation})$$

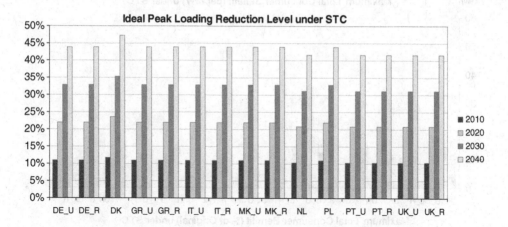

Figure 7.12   Ideal peak load reduction credit, STC

Similar to voltage regulation, peak reduction depends heavily on the location and the installed capacity of dispatchable microsources (Figure 7.12). The effect of reactive power control on line currents seems to be even smaller than its effect on voltage regulation.

7.3.4   Economic Benefits

7.3.4.1   Economic Benefits of Consumers

Local consumer benefit (end consumer perspective) refers to the theoretical room for reduction of retail prices. It is based on the assumption that all economic benefits of a microgrid will be enjoyed by the end consumers, which is most likely to happen under the prosumer consortium ownership model, described in Section 1.5.2, where consumers own and operate multiple microsource units as an aggregated prosumer entity. The benefit can be defined in either relative (per unit) or absolute terms (€/kWh) as follows:

local consumer benefit (p.u.) = 1 − (ideal microgrid consumer electricity price)/(average price from retail market)
local consumer benefit (€/MWh) = (average price from retail market) − (ideal consumer electricity price)

Figure 7.13 shows the maximum total economic benefit of the consumers side in both absolute and relative terms. The economic benefits are the sum of the potential economic benefit obtained from the possibility to choose from more energy suppliers due to market operation, which is termed as market selectivity value, and due to RES proximity to the consumer side, which is termed as locality value. These benefits are calculated under standard test conditions.

The consumer-side selectivity benefits are based on the assumption that all unsubsidized microsource units are operated under zero profit (e.g. when end consumers own the microsource units) and the DSO does not impose any distribution network usage charge for the energy consumption within the microgrid. Comparatively large variations of load-side selectivity

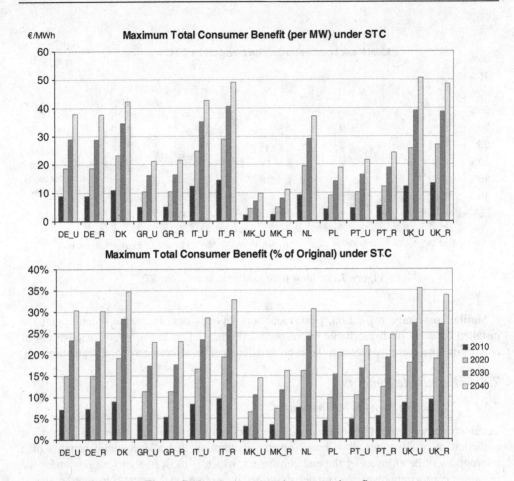

Figure 7.13 Maximum total consumer benefit

benefits have been observed in the various cases due to the different FLH values of dispatchable microsource units and the profit margins between retail prices and microsource costs. As essentially FLH of microsource units is also determined by market prices, the consumer-side selectivity benefit level can be seen to be extremely sensitive to electricity market prices.

Unlike benefits obtained from market selectivity, RES locality benefit is totally dependent on their financial support. This means by default, that market forces will not provide such benefits to consumers due to natural cash flows within a microgrid. If a microgrid comprises only subsidized RES units that are priority dispatched, then consumers will not enjoy any selectivity benefit and the RES locality value – avoided use of system (UoS) charge – will be the only benefit for them. As consumer-side RES locality benefits are deduced from the standard grid charges, UoS charge tariffs determine the national variations in locality benefit value.

Overall, cost saving potentials range from 0.002 Ct/kWh in 2010 to 0.05 Ct/kWh in 2040. The majority of results indicate cost savings ranging from 7% ± 5% in 2010 to 25% ± 10% in 2040. These savings do not include taxes incurred in retail prices that vary widely among the examined countries.

It should be noted that consumers will enjoy the above benefits only with market policy 1 (Section 2.7.3), while with market policy 2, they will have to pay the same electricity prices as without microgrids.

7.3.4.2 Economic Benefits of Microsource Owners

Microsource benefit (microsource perspective) mainly refers to the maximum potential of increasing the selling price under a microgrid setting. This benefit is of course based on the assumption that consumers and the microgrid operator do not benefit economically from the microgrid operation. It can be defined either as a relative value (%) or an absolute value (€/kWh):

> microsource benefit (p.u.) = 1 – (average wholesale price/average price for aggregated microgrid operation)
> micro source benefit (€/kWh) = (average price for aggregated microgrid operation) – (average wholesale price)

Similar to the total local consumer benefit, the total microsource benefit is composed of the locality benefit and the selectivity benefit. Figure 7.14 shows the maximum total benefit on the dispatchable microsource side in both absolute (€/MWh) and relative (%) terms. Variations in microsource costs from 2010 to 2040 basically have negligible impacts on total benefit levels of all examined countries – this can be explained by looking at the maximum total microsource benefit simply as the difference between average retail and wholesale prices in an examined country, which stays largely constant despite microsource cost variation and market price changes. This will be discussed further in the sensitivity analysis of Section 7.4.

Microsource locality benefit is the difference between microsource cost and wholesale price level. It is actually an indication of "opportune operation loss" if the microsource units are forced to sell to the wholesale market instead of local consumers. None of the tested microgrid scenarios under a mid-level price and a real-time pricing setting could provide a direct microsource profit by selling directly to wholesale market.

The selectivity value on the microsource side equals the benefit obtained by consumer-side selectivity, that is incurred by energy consumption from local sources. To reach the full extent of selectivity benefit, consumers are expected to pay the same tariffs as before the microgrid implementation, and the DSO is not expected to charge locally consumed microsource energy for the grid usage. The selectivity benefit approximately yields the maximum achievable profit margin for the units. There is a very close correlation between microsource profitability and retail market price level, enabling high profits at 60–70 €/MWh for high-price countries (e.g. Italy and UK) and much lower results around 20 €/MWh for low-price countries (e.g. Greece and FYROM).

7.3.5 Environmental Benefits

The total emission reduction (regulator perspective) index is the reduction of GHG emission level per kWh consumption in a microgrid when compared to the passive grid operation. It can be defined either as relative value (%) or absolute value (€/kWh):

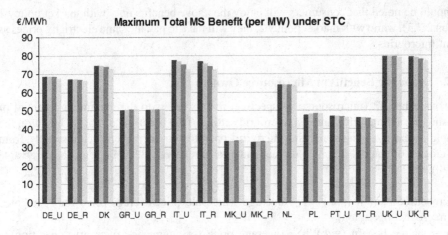

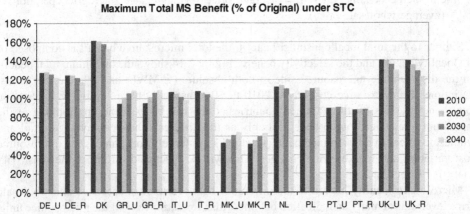

Figure 7.14 Maximum total microsource benefit, STC

total emission reduction (p.u.) = 1 − (GHG emission level in microgrid)/(national GHG emission level, passive case)

total emission reduction (kg/kWh) = (national GHG emission level, passive case) − (GHG emission level in microgrid)

Although SO_2, NO_X and particle matter emission reductions are also expected from microgrid operation, there are currently no explicit trading platforms within Europe as the European ETS market, thus reduction effects under these criteria can be viewed largely as by-products of GHG emission control.

Figure 7.15 demonstrates the environmental benefits of microgrids in absolute and relative values. Microgrid GHG emission levels generally converge to around 200 kg (CO_2 equivalent)/MWh by 2040, despite very different starting points in 2010 – this convergence can be explained by the high resemblance of load-side self-sufficiency level in 2040. GHG reduction credits of microgrid are represented as a percentage of the original emission levels of each examined country. It is quite obvious that countries starting with high emission levels could

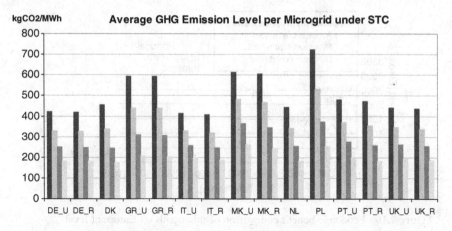

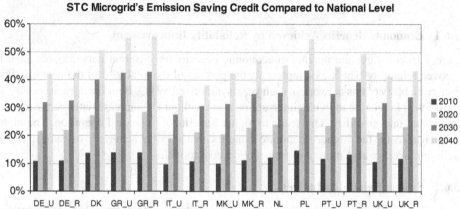

Figure 7.15   Average microgrid GHG emission level and emission saving credit, STC

expect reduction credits as high as over 50% (e.g. Greece and Poland), while countries with lower initial emission (e.g. Italy) enjoy comparatively smaller credits by 2040.

7.3.6 Reliability Improvement

With sufficient local generation and storage capacity, a microgrid can operate in islanded (off-grid) mode serving either the entire load or part of the load, namely the critical loads. The technical requirements of the power electronic interfaces of the DER to enable effective transition from interconnected to islanded operation and to sustain operation of a microgrid in islanded mode were discussed in Chapter 3.

An optimum level of reliability is defined mainly during the planning phase of microgrids, when costs for additional investments and operation are compared to avoided outage costs. Depending on the interruption costs and the given reliability of supply for networks without any microsources, different microgrid reliability benefits can be achieved at European level.

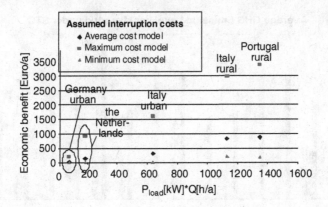

Figure 7.16   Economic benefit comparison of microgrids on European level

7.3.6.1   Economic Benefits Achieved by Reliability Improvement

Figure 7.16 compares the maximum economic benefits for different networks obtained by improved reliability. The x-axis represents the product of the total load of the network times the unavailability of this network in each year, symbolized by PQ. Benefits in each country are almost linear in relation to PQ, as interruption costs without DG increase with increasing total demand and unavailability. This leads to higher benefits of microgrid operation. The higher the outage costs, the higher economic benefits that can be achieved, as shown by the maximum, average and minimum interruption cost model.

7.3.6.2   System Reliability Indices

A reduction of system unavailability Q, by the installation of microgrids that enable (partial) islanded operation, is demonstrated in Figure 7.17 for selected European networks, compared to the case without microsources.

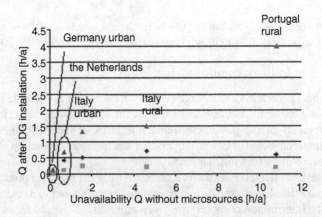

Figure 7.17   System unavailability comparison of different countries

As expected, networks with lower reliability achieve higher improvements by the operation of microgrids, than networks of higher reliability. For instance, in the Portuguese rural network the system unavailability decreases from more than 10 h/a to below 1 h/a with the maximum and average cost model; with the minimum cost model, however, the annual unavailability is reduced to approximately 4 h/a. The improvement in the German urban network and the Dutch network, which have good reliability without microsources, are not significant, although the system reliability is also improved to a certain extent in both networks.

Thus, from the reliability point of view, microgrid operation is most beneficial in networks with lower power quality or for customer segments with comparatively high outage costs.

7.3.6.3 Optimum DG Penetration Level

The optimum DG penetration level is defined as the share of the local DG units required to achieve a minimum number of supply interruptions. As expected, the higher the DG penetration level, the lower the interruption frequency. Specifically, the function of the optimal DG penetration level from reliability perspective is almost linear to DG installed capacity (Figure 7.18). This relationship is important for system planning; that is, as the system interruption frequency without DG istallation is generally known, the DSO is able to decide the optimum DG penetration level from a reliability point of view.

7.3.6.4 Optimum Microsource Location from Reliability Perspective

Investigations show that when only failures caused at the LV network are considered, the optimal microsource location should take into account the following criteria:

- distributed in different protection zones
- located most downstream in the network
- connected close to the load with higher demand
- prioritized to supply sensitive loads

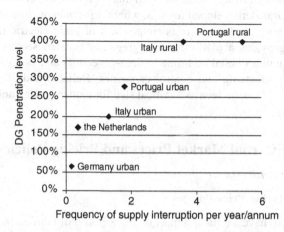

Figure 7.18 Optimal DG penetration for minimum interruption frequency considering average interruption costs

When failures at the MV and HV level are also considered, different microsource locations have the same effect on the frequency of load interruption; in this case, microsources improve system reliability regardless of their location in the LV grid.

7.3.7 Social Aspects of Microgrid Deployment

Three major social benefits of microgrids can be identified as follows:

7.3.7.1 Raise Public Awareness and Foster Incentives for Energy Saving and Emission Cutting

Depending on the microgrid model, DER-created values are shared among end consumers of a network. Residential and commercial participants of a microgrid will likely enjoy better electricity tariffs and can even be rewarded by incentive programs if they own RES microsources or participate in demand-side integration (DSI) programmes. This type of economic signal (efficiency values created within a microgrid) can be seen as a strong driving force for acceptance and promotion of microgrids.

7.3.7.2 Creation of New Research and Job Opportunities

Microgrid implementation requires knowledge, expertise and customized hardware and software solutions that are currently not directly available for both supply- and demand-side players in the market. Opportunities for new research posts and job openings will be offered for technology providers, device manufactures, research institutes, and so on. Opportunities will become available not only in the design (stereotyping) and installation (standardization) stages, but also in daily operation and maintenance of a microgrid, for example requirements for real-time metering, communication and control both within and outside of microgrids.

7.3.7.3 Electrification of Remote or Underdeveloped Areas

For a long time, it has been widely acknowledged that DER units are extremely well suited for electrifying remote or underdeveloped regions, where it is not economic to interconnect to a nearby distribution grid or where there is simply lack of basic electric infrastructure. The microgrid concept provides a platform for aggregating isolated sectors of self-sufficient households or communities based on limited micro-generators and storage units into a more robust network with balancing and control capacities. Reliable and affordable supply of electricity via microgrid can be seen as a critical step for modernization and industrialization of the local economy.

7.4 Impact of External Market Prices and Pricing Policies

7.4.1 Sensitivity Analysis

7.4.1.1 External Market Price Policies

External market prices have a critical impact on microgrid profitability. In Figure 7.19, three potential price settings are assumed, as inputs to the sensitivity analysis of microgrid profitability (see also Section 1.5.3 and Table 1.1):

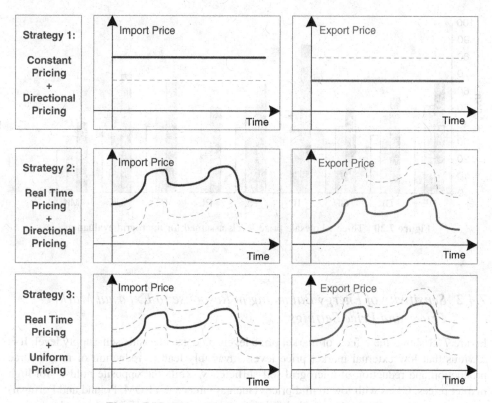

Figure 7.19 External price policies applied to microgrids

- Pricing policy 1 (fixed prices): directional + constant pricing: economic island model
- Pricing policy 2 (flexible prices): directional + variable (ToU) pricing: hybrid model
- Pricing policy 3 (uniform prices): uniform + variable (ToU) pricing: exchange model.

7.4.1.2 Wholesale Market Price Levels

Due to the extremely volatile nature of wholesale electricity prices, it is not clear whether future electricity prices will increase (e.g. due to scarcity of conventional resources) or decrease (decreasing demand due to higher efficiency) compared to their current level. Therefore, electricity price levels (annual average of wholesale price in all countries) are assumed to be constant, irrespective of time, and three general price levels are assumed for all countries and for all scenarios considered. Figure 7.20 shows the annual average values of high, mid and low prices at wholesale level, respectively, as annual average values for the analyzed countries. The high price scenarios are taken as equal to the peak prices from the past decade, while mid and low cases are respectively calculated as 80% and 60% of peak value.

In order to ensure microgrid profitability under low market prices, market acceptance of local consumption satisfaction is assumed here for all cases.

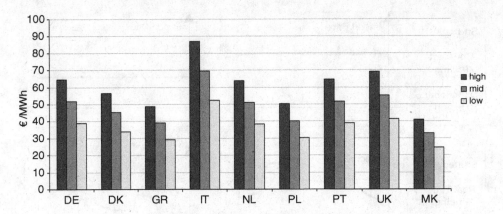

Figure 7.20 Three wholesale price levels assumed for microgrid evaluation

7.4.2 Sensitivity of Energy Balancing in Response to External Market Prices and Price Settings

Figure 7.21 shows the effects of market price levels on consumer-side self-supply level. It is obvious that low external market price levels invariably lead to reduction of microsource production and reduction of microgrid self-sufficiency, while the opposite holds – for high market prices. Cases with low to mid prices (such as Greece, FYROM, Poland and Portugal) are much more sensitive to annual wholesale price changes (up to ±15%) compared to higher-price cases (largely below ±5%).

Compared to real-time pricing, application of fixed prices has a negligible impact on higher-price cases, but it visibly increases self-sufficiency level of low- to mid-price cases. Recognition of locational value (uniform pricing), however, always leads to lower self-sufficiency levels due to a general reduction of retail purchase tariffs.

Figure 7.22 shows the effects of external market price changes on the microsource side of self-supply level.

In contrast to the consumer-side self-supply level, the microsource contribution to self-supply increases with lower market price levels and decreases as average market prices increase. This can be easily explained as prices determine export opportunities during peak price hours. In a number of cases, microsource contribution to self-supply level varies by ±4%, while in most of them this falls within ±1%, which indicates a very small impact of market prices on microsource export opportunities. This suggests that export opportunities, even with a high penetration of local DER (2040 case) will be primarily determined by the load level instead of the external price level. Fixed pricing has similar effects as it leads to market prices variations generally within ±2%. Uniform pricing, however, results in considerably larger reduction of microsource self-supply, thus increasing exports to the upstream network by 10–15% in 2030 and 2040 scenarios.

The effect of market price levels and pricing policies on microsource dispatch, expressed as mean FLH, can be seen in Figure 7.23. The impact of market price level on microsource FLH shows a very similar trend to that of the consumer-side self-supply index shown in

Figure 7.21 Market price and pricing policy impact on consumer-side self-supply

Figure 7.21. FLH of low-price cases exhibits higher sensitivity to average wholesale price variations (up to ±60% in the case of FYROM), while high-price cases (e.g. Germany, Denmark, Italy, the Netherland and the UK) generally experience variations less than ±10% and are hardly impacted by wholesale market price levels.

Fixed pricing, as defined in Section 1.5.3, has negligible effects on high-price cases, but can increase DER FLH by as much as 80% compared to real time (ToU) pricing for low-price cases, such as FYROM, Greece and Poland. This can be understood as the benefit of price stability in adverse market settings. Uniform pricing provides a 20% to 80% FLH reduction in 2010, which is decreased drastically with time; for example, in some cases FLH starts to increase by 2040 as a result of uniform pricing.

The effects of market price levels and pricing policies on the required premium support level for RES can be seen in Figure 7.24. The support levels are calculated as the difference between mean RES production cost and average retail prices. Owing to the linear correlation between market prices and required RES support, variations in average wholesale prices level induces proportional changes in required RES support, that is higher prices lead to reduction of required support and vice versa. In general, a ±25% market price variations leads to a change in required RES support by 10–20%.

Figure 7.22 Market price and pricing policy impact on microsource contribution to self-supply

Fixed (constant) pricing leads to an increase in the required premium support for some countries (e.g. DE, IT, NL and PT) and lower support for others (e.g. DK, PL and the UK). Uniform pricing, on the other hand, is assumed not to interfere with RES support and consequently has no impact on required national support levels.

7.4.3 Sensitivity of Technical Benefits (Losses) in Response to External Market Prices and Price Settings

Figure 7.25 shows the impacts of market price levels and price settings on losses. The trends are similar to those of the load-side self-sufficiency level index of Figure 7.21, since losses are roughly dependent on the levels of microsource exports for all examined simulation cases.

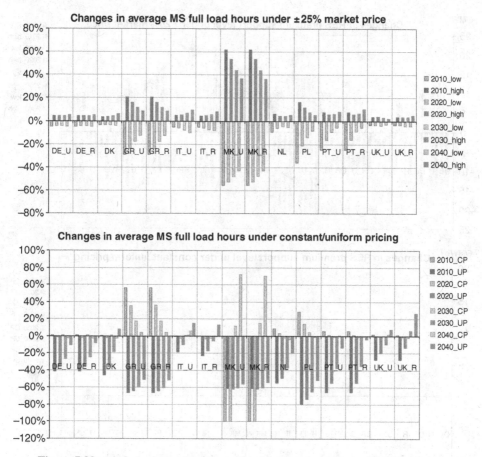

Figure 7.23   Market price and pricing policy impact on microsource full load hours

7.4.4 Sensitivity of Economic Benefits in Response to External Market Prices and Price Settings

7.4.4.1 Consumer Side

The effects of market price variations and different pricing policies on the total economic benefit on the consumer side are shown in Figure 7.26. Differences between the various cases under the same wholesale price variations are comparatively small, indicating a benefit variation ±20% to ±40% for a ±25% variation of market prices. Fixed pricing has mostly negative impacts on maximum consumer benefit, with small benefit reductions under 10% for most cases except FYROM, where a benefit reduction as high as 35% is calculated due to a drastic reduction of microsource full load hours (even to zero). Uniform pricing could significantly raise the maximum total consumer benefit by 2–6 times initially in 2010, but this increase is quickly damped with time.

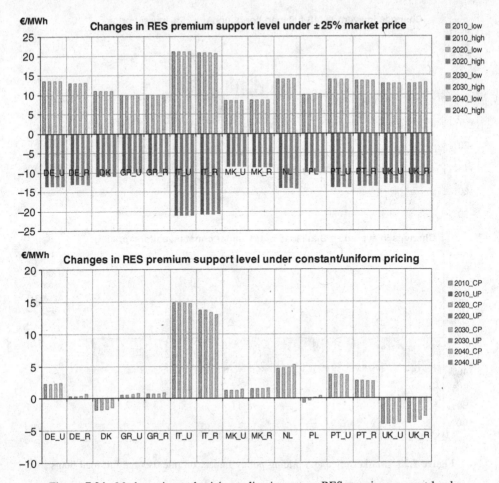

Figure 7.24 Market price and pricing policy impact on RES premium support level

7.4.4.2 Microsource Side

The potential effects of market price variations and pricing policies on the maximum total benefit of microsources are shown in Figure 7.27. The total microsource benefit is almost solely determined by the difference between the selling price and the average wholesale price in the market. Wholesale market price levels and fixed pricing schemes have basically negligible impacts (below 5%) on its value except the case of FYROM. Uniform pricing, however, is potentially capable of changing the mean microsource selling price and can therefore lead to total microsource benefit changes as high as ±50% to ±100%.

7.4.5 Sensitivity of Environmental Benefits in Response to Market Prices and Price Settings

The effects of market price levels and pricing policies on microgrid emission savings are shown in Figure 7.28. There is a close correlation between microgrids self-sufficiency level

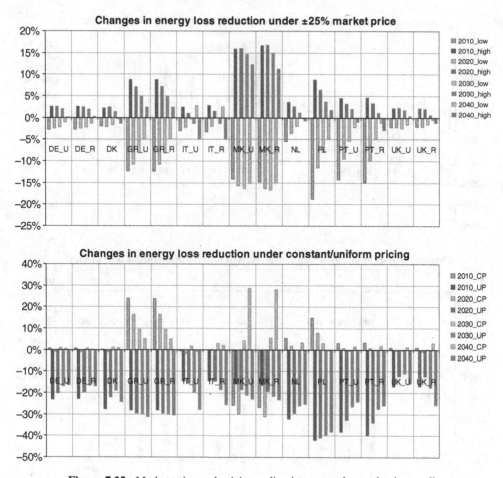

Figure 7.25   Market price and pricing policy impact on loss reduction credit

and its emission reduction credit. There are different national emission level baselines and large variations of emission levels in response to external prices and pricing settings, but the actual impact of both criteria is comparatively small (less than ±10%). The only exception is the effect of uniform pricing, which could potentially lead to a reduction of more than 25% in terms of GHG savings compared to real-time pricing.

7.5 Impact of Microgrid Operation Strategy

Figure 7.29 compares the maximum total consumer benefit obtained from the combined operation mode (the objective function is the summarized total cost, and all technical constraints are applied) used as a reference to the following three individual operating strategies for one sample microgrid (UK urban) in 2010 to 2040 scenarios:

- Economic strategy: objective function is the maximization of profits as seen from microsource aggregator perspective; no network technical constraints are applied.

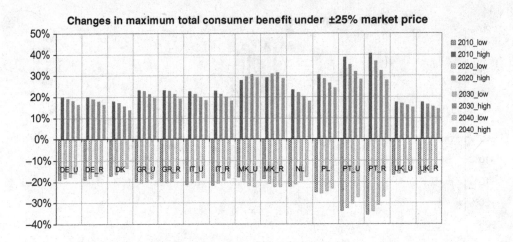

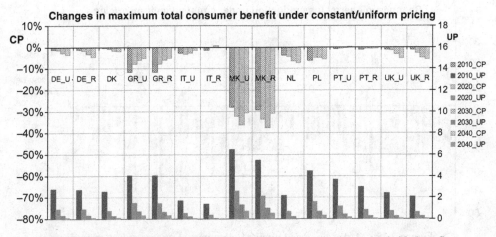

Figure 7.26   Market price and pricing policy impact on maximum load-side benefit

- Technical strategy: objective function is the minimization of losses; all technical network constraints are applied as seen from a DSO perspective.
- Environmental strategy: objective function is the minimization of emissions from a regulator perspective; no network technical constraints are applied.

It can be seen that the differences between the strategies are not significant when the self-supply level of a microgrid is below 50%.

As expected, multi-objective optimization provides a balanced set of benefits, while the single-objective optimizations provide higher benefits for the respective individual benefit at the expense of the other benefits.

In order to facilitate illustration, the following primary indices are defined:

economic index = maximum profit per kWh sold energy
technical index = annual grid energy losses in kWh per year
environmental index = avoided GHG emission in CO_2 equivalent per kWh

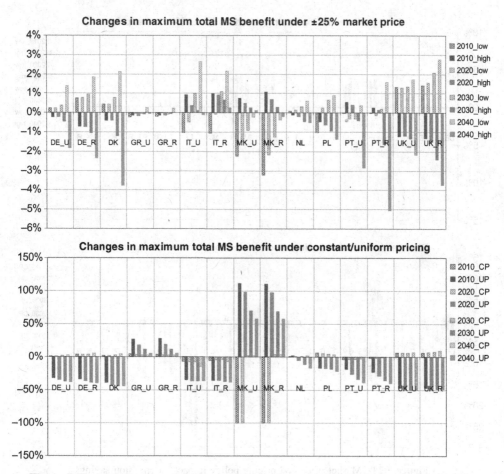

Figure 7.27 Market price and pricing policy impact on maximum microsource side benefit

These indices are calculated for each of the three strategies and compared to the corresponding indices obtained from the combined multi-objective optimization for the studied European microgrids. The relative differences are plotted respectively in Figures 7.30–7.32 as percentage values where the multi-objective calculation results are taken as reference (i.e. the denominator).

It can be clearly seen that the combined multi-objective optimization provides a balanced satisfaction of all objectives and can be physically viewed as a compromise between the different microgrid stakeholders in order to arrive at the global optimum of economic, technical and environmental performances.

In Figures 7.33 and Figure 7.34, the voltage and loading reduction indices are compared for all examined cases. It can be seen that the combined strategy achieves the same level of network performance as the technical optimization strategy, while the economic and environmental optimization strategies fail to do so.

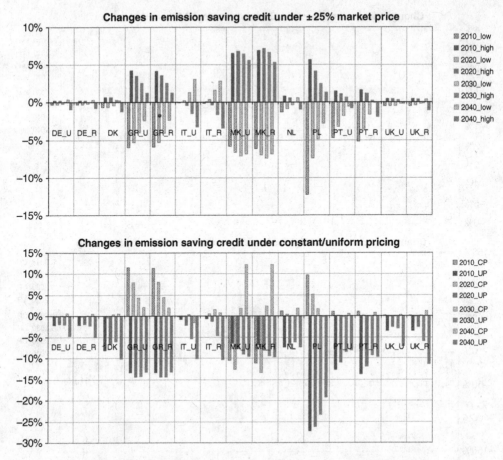

Figure 7.28 Market price and pricing policy impact on emission savings

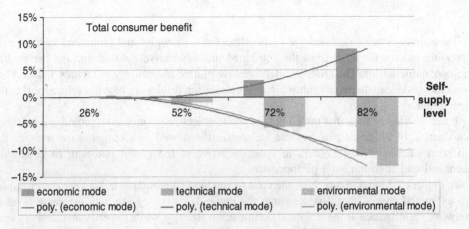

Figure 7.29 Sample microgrid operation mode impact over different self-supply levels

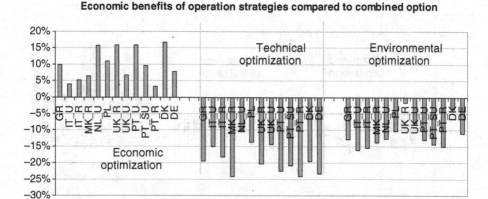

Figure 7.30 Operation strategy impact on maximum economic benefits

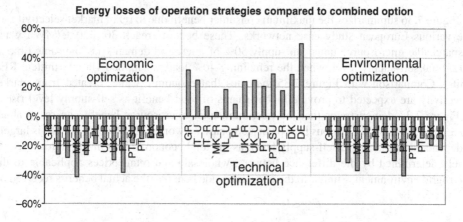

Figure 7.31 Operation strategy impact on energy losses

7.6 Extension to European Scale

As shown in Section 7.2 the microgrid self-supply level is the most important factor affecting the benefits that it provides to the various stakeholders. Figure 7.35 demonstrates the maximum total consumer benefit for the various European study case networks as a function of self-supply level. Benefit increases as the microgrid self-supply level increases and reaches 35 ±25 €/MWh at 90% (consumer-side) self-supply level. This maximum benefit is the sum of potential price reductions due to local trading, as well as network and emission charge reductions. The values shown are based on the assumption that all economic benefits are obtained by the end consumer. In practice, it is expected that microsources, DSO and suppliers or other potential intermediary parties (e.g. ESCOs) will very likely share this total benefit.

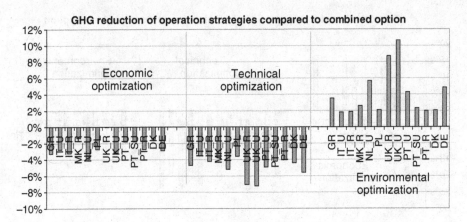

Figure 7.32 Operation strategy impact on GHG reduction credits

Figure 7.36 summarizes the maximum consumer benefit due to retail market selectivity for the various European study case networks. These benefits reach 30 ±20 €/MWh when dispatchable microsource units can supply 50% of microgrid demand (i.e. the same time as 90% total self-supply level, when the remaining 40% load is supplied via intermittent RES units). Comparison with Figure 7.35 indicates that economic values created from market selectivity are expected to provide higher shares of total benefit as self-supply level rises.

Figure 7.37 demonstrates that the maximum total microsource benefit can reach about 60 ±30 €/MWh for the various European study case networks. This level of benefit is largely independent of microgrid self-supply level, as the total economic benefit of microsources is solely determined by the difference between wholesale and retail prices applicable to the microgrid environment. Similar to the consumer-side benefits, the actual benefits received by

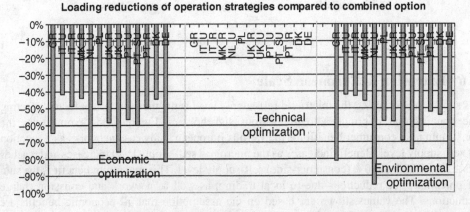

Figure 7.33 Operation strategy impact on loading reduction

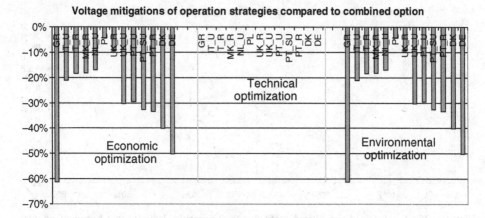

Figure 7.34 Operation strategy impact on voltage variation

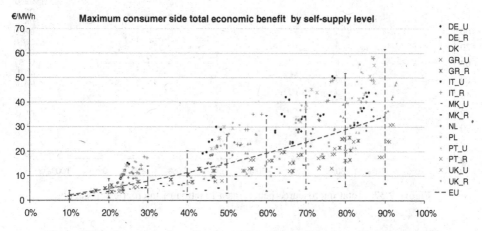

Figure 7.35 Summarized total consumer benefit as a function of load self-supply

the microsource owners are likely to be smaller, since they can be potentially shared with other involved parties.

Figure 7.38 summarizes the ideal loss reductions for the various European study case networks. A reduction of about 75% ±20% can be expected at 90% self-supply level.

Figure 7.39 summarizes the ideal voltage regulation improvement for the various European study case networks. A reduction of 50% ±15% can be expected at 90% self-supply level.

Moreover, a reduction of about 40% ±12% of the peak load at 90% self-supply level (Figure 7.40) and a reduction of GHG emissions by about 55% ±25% at 90% self-supply level (Figure 7.41) can be achieved under an ideal setting.

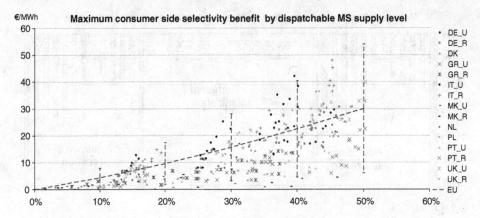

Figure 7.36 Summarized consumer benefit due to microgrid selectivity in a European context

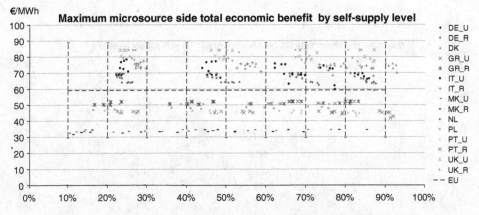

Figure 7.37 Total microsource benefit as a function of microgrid self-supply

7.7 Conclusions

Network operation under a microgrid paradigm provides novel possibilities for overcoming the conflicting interests of different stakeholders and for achieving a global socio-economic optimum in the operation of distributed energy sources. Economic, technical and environmental impacts of microgrids are closely related to the operating decisions about the outcomes of microsources, storage units and DSI programmes. Proper planning of a microgrid requires knowledge and simulation of its potential operating conditions, while at the same time different planning decisions; especially those related to DG/RES penetration level, will critically affect potential benefits of a microgrid.

In this chapter, microgrid benefits have been identified for a number of representative distribution grids using a variety of input data from representative European countries and

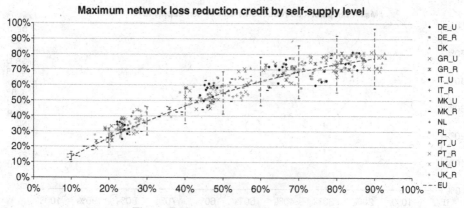

Figure 7.38   Summarized loss reductions

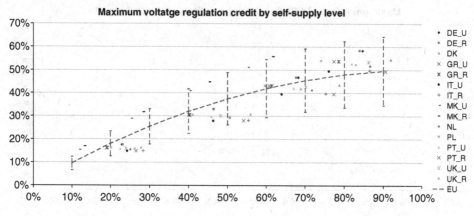

Figure 7.39   Voltage regulation

regions. In order to quantify the benefits obtained from different real-time operating conditions, each case study network has been simulated as a consecutive day-to-day scheduling problem with annual sets of stochastic weather, market and demand data. A multi-objective optimization algorithm has been used that incorporates technical and environmental aspects either as economic objectives or as operational constraints. An optimal power flow technique is used in combination with meta-heuristic methods to estimate real-time system states. Finally, reliability improvements and social benefits have been considered independently from daily microgrid operation and are thus studied separately.

Study results indicate that broad ranges exist for literally all economic, technical and environmental benefits potentially obtained from microgrids in different countries and regions. Considering the high uncertainties in the underlying assumptions, it is very difficult to provide reliable, quantified benefits of satisfactory precision. Despite individual differences, however,

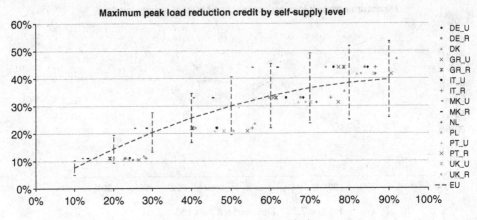

Figure 7.40 Peak load reduction

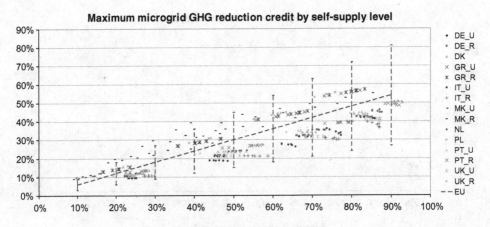

Figure 7.41 GHG emission reduction

general trends of microgrid effects can be safely observed. Statistical analysis of the calculated results also suggests that the energy self-sufficiency level (demand-side) is a good indicator of the potential values of most benefit indices.

The main results of the benefits studies are as follows:

1. Microgrids can be **profitable** to invest in and to operate, given the current market situation in Europe. However, a suitable regulatory framework, including proper policy and financial support, needs to be available.
2. Microgrids offer a **local market** opportunity for over-the-grid energy trading between microsources and end consumers.
3. Microgrids can **maximize the total system efficiency** as they represent the interests of microsources, end consumers and the local LV grid as a whole.

4. Microgrids allow for real-time, **multi-objective dispatch** optimization to achieve economic, technical and environmental aims at the same time.
5. Microgrids can accommodate different ownership models and provide **end consumer motivation** where other concepts fail to do so.

Benefits of Microgrid Operation

A microgrid could potentially offer **economic benefits** (single or multiple from list) such as:

- price reduction for end consumers,
- increased revenues for microsources,
- investment deferral for distribution system operators.

Economic benefits arise mainly from locality values created by local retail market and selectivity values created by optimized real-time dispatch decisions.

To achieve the expected economic benefits, it is necessary to recognize local (over-the-grid) energy trading within a microgrid, to apply real-time import and export prices for microgrids and to have an (optimal) RES support scheme and favorable tariffs.

A microgrid could potentially offer **technical benefits** (single or multiple from list) such as:

- energy loss reduction,
- lower voltage variation,
- peak loading (congestion) relief,
- reliability improvement.

Technical benefits can be either traded in a local service market between microsource and DSO or implemented as price signals. To achieve the expected technical benefits, optimal dimensioning and allocation of microsources as well as coordinated multi-unit microsource dispatch, based on real-time grid conditions is necessary.

Environmental benefits of microgrid can be mainly attributed to:

- shift toward renewable or low-emission fuels used by internal microsources,
- adoption of more energy-efficient technologies, such as CHP,
- adoption of DSI options.

Social benefits of microgrids can be summarized as:

- raising public awareness and fostering incentive for energy saving and GHG emission reduction,
- creation of new research and job opportunities,
- electrification of remote or underdeveloped areas.

Reference

1. More Microgrids (2006–2009) Advanced Architectures and Control Concepts for More Microgrids, FP6 STREP, Contract no.: PL019864., http://www.microgrids.eu.

Index